AF602052

Consumer Oriented Agricultural Marketing

The Author

Dr. Ruchira Shukla is presently working as Associate Professor (Agribusiness Management) at the ASPEE Agribusiness Management Institute, Navsari Agricultural University, Navsari, Gujarat, India. She has more than 13 years of teaching experience at UG and PG level. She is actively engaged in teaching of various subjects of Agribusiness Management such as marketing management, agribusiness management, agricultural marketing, rural marketing, sales and distribution management, food retail management, entrepreneurship development and various other management and marketing related subjects. She is also associated with many research projects on various aspects of marketing and value addition of agro produce. She has authored 3 books, written more than 40 research papers, 9 chapters in books and around 60 popular articles.

Consumer Oriented Agricultural Marketing

– *Author* –

Ruchira Shukla

Associate Professor (Agribusiness Management)
ASPEE Agribusiness Management Institute
Navsari Agricultural University, Navsari (Gujarat)

2017

Scholars World

A Division of

Astral International Pvt. Ltd.

New Delhi – 110 002

Publisher's Note:

Every possible effort has been made to ensure that the information contained in this book is accurate at the time of going to press, and the publisher and author cannot accept responsibility for any errors or omissions, however caused. No responsibility for loss or damage occasioned to any person acting, or refraining from action, as a result of the material in this publication can be accepted by the editor, the publisher or the author. The Publisher is not associated with any product or vendor mentioned in the book. The contents of this work are intended to further general scientific research, understanding and discussion only. Readers should consult with a specialist where appropriate.

Every effort has been made to trace the owners of copyright material used in this book, if any. The author and the publisher will be grateful for any omission brought to their notice for acknowledgement in the future editions of the book.

Cataloging in Publication Data--DK
Courtesy: D.K. Agencies (P) Ltd. <docinfo@dkagencies.com>

Shukla, Ruchira, author.
Consumer oriented agricultural marketing / author, Ruchira Shukla.
pages cm
Includes index.
ISBN 9789387057401 (International Edition)

1. Farm produce--India--Marketing. 2. Consumer behavior--India. I. Title.

HD9016.I42S58 2017 DDC 381.410954 23

Published by : **Scholars World**
A Division of
Astral International Pvt. Ltd.
– ISO 9001:2015 Certified Company –
4736/23, Ansari Road, Darya Ganj
New Delhi-110 002
Ph. 011-4354 9197, 2327 8134
E-mail: info@astralint.com
Website: www.astralint.com

Preface

Agricultural marketing covers the services involved in moving an agricultural product from the farm to the consumer. Numerous interconnected activities are involved in doing this, such as planning production, growing and harvesting, grading, packing, transport, storage, agro- and food processing, distribution, advertising and sale. Marketing systems are dynamic; they are competitive and involve continuous change and improvement. Businesses that have lower costs, are more efficient, and can deliver quality products, are those that prosper. Those that have high costs, fail to adapt to changes in market demand and provide poorer quality are often forced out of business. Marketing has to be customer-oriented and has to provide the farmer, transporter, trader, processor and all involved in agri value chain with a profit. This requires those involved in marketing chains to understand buyer requirements, both in terms of product and business conditions.

New marketing linkages between agribusiness, retailers and farmers are gradually being developed, e.g. through contract farming, group marketing and other forms of collective action. Increasing attention is being paid by government, NGO's and Private players to ways of promoting direct linkages between farmers and buyers within a value chain context.Farmers frequently consider marketing as being their major problem. However, while they are able to identify such problems as poor prices, lack of transport and high post-harvest losses, they are often poorly equipped to identify potential solutions. Successful marketing requires learning new skills, new techniques and new ways of obtaining information.

Considering above Agribusiness Management, Agricultural Marketing, Agri input Marketing and related courses are being taught in various under graduate and post graduate programmes in Agriculture and allied subject. My prime objective is to ensure that the book helps the students and practioners to learn and practice the various aspects of Agricultural Marketing with a consumer and market orientation which is the need of the hour in Agriculture produce marketing. This book focuses on Four "P" Approach and consumer orientation as applied to agricultural marketing.

I have tried to collect and compile useful information about this very important subject for the student and practioners.

I would like to express my gratitude to all those who provided support, talked things over, read, wrote, offered comments and assisted in the editing, proofreading and design.

I am grateful to Honourable Vice Chancellor of Navsari Agricultural University for providing constant encouragement, motivation and inspiration for carrying out useful work especially in the field of marketing. I am heartily grateful to the Directorate of Research at Navsari Agricultural University for providing constant encouragement which helped in developing a confidence which was much needed. I am grateful to the Dean of the Institute for providing for providing valuable suggestions and constant encouragement during the preparation of the manuscript.

My Colleagues at ASPEE Agribusiness Management Institute have provided necessary support, encouragement and constructive suggestions for the preparation of manuscript which is gratefully acknowledged. I am thankful to my students in the classroom whose queries and curiosities have helped me better understand their requirements.

Above all I would like to thank my husband Dr. Abhishek Shukla for all the inspiration and motivation and boosting my confidence for the project. I also thank my wonderful children: Shourya and Abhishree who supported and encouraged me in spite of all the time it took me away from them.Words fall short for my Parents including my in-laws, who have always inspired and supported me throughout my career and authoring this book and I really appreciate it. My deepest thanks to all my well wishers and friends who have always appreciated and encouraged my efforts.

Lastly I shall be glad to receive constructive comments, suggestions and corrections from seniors, colleagues and students.

Ruchira Shukla

Contents

List of Figures

List of Tables

Chapter 1

Introduction to Agricultural Marketing

WHAT IS MARKETING?

There are many definitions of "marketing."

'Marketing is a social and managerial process by which individuals and groups obtain what they want and need through creating, offering and exchanging products of value with others'

– *Philip Kotler*

The American Marketing Association (AMA) uses the following: "The process of planning and executing the conception, pricing, promotion, and distribution of ideas, goods, and services to create exchanges that satisfy individual and organizational objectives."

From this definition, we understand that:

- ☆ Marketing is an ongoing process. The environment is "dynamic"—what customers want today is not necessarily what they want tomorrow.
- ☆ Marketing involves both planning and implementing (executing) the plan.

Some of the main issues involved include:

- ☆ Marketers help design products based on what customers want and what can be made available with technology and price being constraint at any given time.
- ☆ Marketers distribute products—there must be some efficient way to get the products from the factory to the end-consumer.

- Marketers promote products. Promotion involves advertising and much more. Other tools to promote products include trade promotion (store sales, coupons, and rebates), obtaining favorable and visible shelf-space, and obtaining favorable press coverage.
- Marketers also price products. We know from economics that higher the price, the lower the quantity demanded. In some cases price may provide the customer with a "signal" of quality. Thus, the marketer needs to price the product to (1) maximize profit and (2) communicate a desired image of the product.
- Marketing is applicable to services and ideas which are intangibles as well as to tangible products.

Here are two particularly relevant to Agricultural marketing.

The first is.

Marketing involves finding out what your customers want and supplying it to them at a profit

This stresses two important points:

- The marketing process has to be **customer oriented**;
- Marketing, a **commercial process**, has to provide farmers, transporters, traders, processors, *etc.* with a profit or they will be enable to stay in business.

Marketing therefore involves:

- Identifying buyers;
- Understanding what they want in terms of products and how they want to be supplied;
- Operating a production-marketing chain that delivers the right products at the right time;
- Making enough profit to continue to operate.

The second useful definition is.

The series of services involved in moving a product from the point of production to the point of consumption

This definition emphasizes that marketing is a series of inter-connected activities. In the case of agricultural marketing these include:

- Planning production;
- Growing and harvesting;
- Grading of products and their packing, transport, storage, processing, distribution and sale;
- Sending information from production area to market (*e.g.* products available, volumes) and from market back to producing areas (*e.g.* prices and supply levels, consumer preferences and changes in taste).

All of these activities are links in the production-marketing chain. Like any chain, it is only as good as its weakest link.

Marketing systems are dynamic. They are competitive and involve continuous change and improvement. Suppliers who have lower costs, are more efficient and can deliver quality products are those who survive and prosper. Those who have high costs, do not adapt to changes in market demand and provide poorer quality are often forced out of business.

AGRICULTURAL MARKETING

Agricultural marketing includes the movement of agricultural produce from farms where it is produced to the consumers or manufacturers. This covers physical handling and transport, initial processing and packing to simplify handling and reduce wastage, grading and quality control to simplify sales transactions and meet different consumers' requirements, and holding over time to match concentrated harvest seasons with the continuing demands of consumers throughout the year.

For the farmer, the strategic function of the marketing system is to offer him a convenient outlet for his produce at a remunerative price. To the consumers and the manufacturers of agricultural raw materials, assurance of a steady supply at a reasonable price is the vital service. Prices are determined through free market process by negotiations at rural purchase, wholesale and retail stages, and represent a balance between the consumers' ability to pay and the farmers' need for incentive to produce.

An effective marketing system will be geared towards expanding the range and types of consumer service, and thus offer procurement outlets.

Agricultural marketing also includes the marketing of production inputs and services to the farmers. Some of these include fertilizers, pesticides and other agricultural chemicals; livestock feed; farm machinery, tools and equipment. As the mass of small farmers becomes aware of the value of these supplies, the organization of distribution systems adapted to their needs becomes vital. Through all the stages of marketing, financing and easy access to credit is vital if goods are to move freely.

HOW CAN IMPROVED MARKETING HELP FARMERS?

Most farmers see themselves as "price takers", thinking that they have no control over prices and have to accept what is offered. They do not always know how to find new buyers nor how market demand is changing and which products are most profitable to grow. They lack the understanding to improve the prices they receive and the profitability of their production.

Farmers need to become better informed about the market. They can then start to make decisions on how to improve their marketing. Farmers should make decisions such as what they should do or what products to sell. Commercial decisions like these must be made by the farmers themselves. Farmers need to become owners of new ideas. In this way they become committed and aware that they are responsible for the success or failure of what they do. This increases the chances of a successful outcome.

When asked about their problems, farmers commonly identify marketing issues as their key constraint. Problems highlighted are usually lack of markets, poor prices, inadequate roads and poor communications.

However, while farmers can usually state clearly their problems they often face difficulties in identifying potential solutions.

Farmers are generally highly skilled in agricultural techniques but marketing requires learning new skills, new techniques and new sources of information. Armed with business and marketing skills farmers will be better able to run their farms profitably.

Small-scale farmers face the biggest marketing problems. Box 1 compares the strengths and weaknesses of large-scale farms and small-scale farms. Small farmers need the most support and that their success depends on getting the best prices possible. This can be done by obtaining better information about marketing and the different marketing options available to them.

Strengths and Weaknesses of Small and Large Farms

Strengths	*Weaknesses*
Small farms	
☆ Cheap family labour is available. Smalls farms are suitable for labour-intensive products (*e.g.* those requiring transplanting, pruning and multiple harvests by hand).	☆ Small farms produce only limited quantities. They are often located far away from major markets. Education standards are often low. Small farmers are reluctant to adopt new technology.
☆ Small farms can grow products that require attention to detail.	☆ They face difficulties in obtaining information, capital and support.
☆ Small farms can effectively supply low-volume specialized niche markets and value-added products, such as herbs, flowers and ornamental plants.They can also supply local processors.	☆ They are weak in negotiation and often lack confidence, especially when dealing with traders and companies.
	☆ Small farmers tend to be averse to risk. They need income stability and cannot afford losses.
Large farms	
☆ Large farms are suitable for mechanized, large-scale production for major crops like wheat, sugar cane *etc.*	☆ Large farmers often have high overhead costs.
☆ They can grow crops that require a large capital investment.	☆ They can be poor at organizing and controlling large numbers of workers.
☆ They are best equipped to produce and sell produce in large volumes to major buyers.	☆ They cannot easily service small and niche markets.
☆ They have access to capital, information and technology.	

WHAT IS THE ROLE OF MARKETING?

Marketing and the Rural Economy

Those who carry out marketing have a strong incentive to increase the value of rural trade, because increased sales should lead to higher profits. Rural businesses include suppliers of inputs, buyers of produce, transporters, storage companies, processors and wholesalers. They can range in size from individual entrepreneurs to large-scale agribusinesses, but whatever their size, all stand to gain from improvements in the marketing process.

Businesses are often said to be exploiting farmers and making unfair profits. They certainly try to maximize their profits, but without such businesses farmers would not be linked to markets and would not be able to sell all their produce. Traders and other rural businesses can help farmers find new markets and lower their costs. All of this leads to improved production opportunities and higher incomes for farmers.

Rural marketing businesses are often small, have limited resources and are traditional in outlook.

Governments can help farmers in many ways, without actually working with them directly. Promotion of competition, provision of market information and improvement of market infrastructure are powerful ways to ensure good returns for farmers.

Marketing and Consumers

Consumers want to pay low prices. Farmers want to receive high prices and to be paid as much of the consumer price as possible. The best way of achieving a balance between these two conflicting aims is through an efficient and low-cost marketing chain. This generally involves using larger scale transport (achieving economies of scale), reducing losses, and reducing other costs.

Consumer preferences for food products are constantly changing and developing, particularly in the case of agricultural and horticultural products. Consumers need a production-marketing chain that can respond to their changing tastes. The marketing system needs to deliver the volumes, quality and variety of safe and nutritious food products that consumers require. It needs to be sufficiently dynamic so that it can continue to provide consumers with choice by developing and delivering new products. Agricultural marketing too needs to be consumer oriented.

MARKETING MANAGEMENT PHILOSOPHIES

There are different philosophies guiding a marketing effort in a organization. These have evolved over a period of time.

Production Concept

One of the oldest concepts (Industrial revolution-seventeenth century to 19th century)."Supply creates its own demand" it holds that consumers will prefer products that are widely available and inexpensive, so emphasis on achieving high production efficiency, low cost and mass distribution. Product features were not given importance.

Product Concept

Consumers will favour those products that offer the most quality, performance or innovative features. The focus is on making superior products."Marketing Myopia" occurs when a marketer is excessively preoccupied with product development, manufacturing or selling and ignores customer needs, wants and interests

Selling Concept

Organization should undertake an aggressive selling and promotion effort. The purpose is to sell more stuff to more people to make more money. Aim is to sell what they make rather than what market wants.

Marketing Concept

Emerged in 1950's where instead of "product centered, make and sell" philosophy emphasisis on a "customer centered, sense and respond" philosophy.

Marketing concept holds that key to achieve organizational goals is to be more effective than competitors in creating, delivering and communicating superior customer value to its chosen target markets. The companies embraced the concept

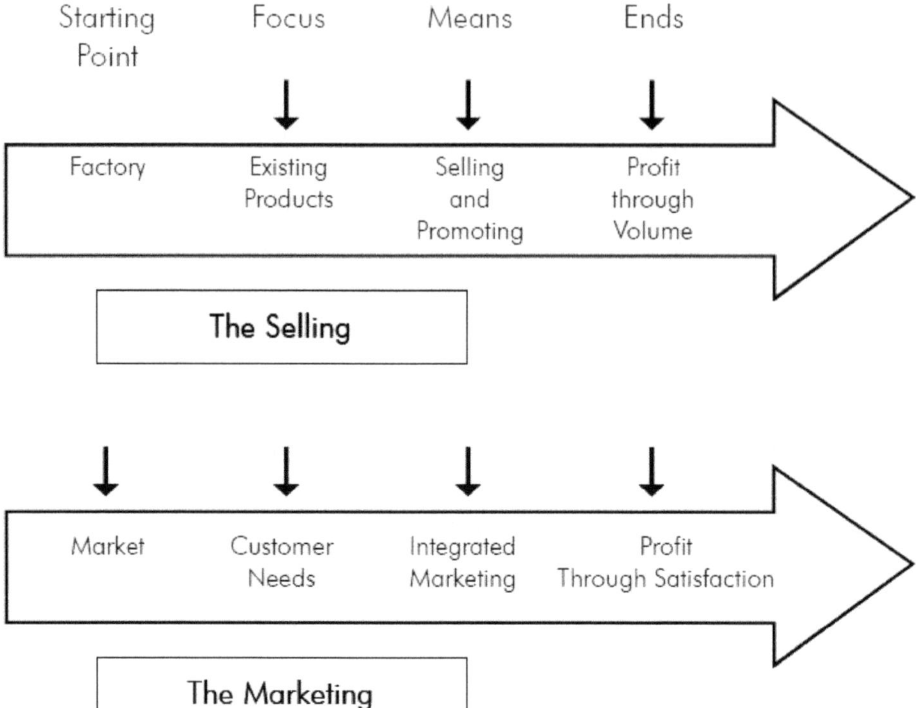

Figure 1.1: Difference between Selling and Marketing Approach.

of coordinated marketing management directed towards twin goals of customer orientation and profitability.

Marketing Concept vs. Selling Concept

Two approaches to marketing exist. The traditional selling concept emphasizes selling existing products. The philosophy here is that if a product is not selling, more aggressive measures must be taken to sell it *e.g.*, cutting price, advertising more or hiring more and aggressive sales-persons.

The marketing concept, in contrast, focuses on getting consumers what they seek, regardless of whether this entails coming up with entirely new products.

Societal Marketing Concept

Marketers carefully consider the role they are playing in terms of social welfare. The societal marketing concept calls upon marketers to build social and ethical considerations into their marketing practices. They should balance the often conflicting interests of company profits, consumer want satisfaction and public interest. It believes that companies that act in socially responsible manner gain goodwill, apart from reaping profits.

Table 1.1: Marketing Management Philosophies

Concepts	*Meaning and Implications*
Production Concept	Consumers favor products that are available and highly affordable Improve production and distribution
Product Concept	Consumers favor products that offer themost quality, performance, and innovative features
Selling Concept	Consumers will buy products only if thecompany promotes/sells these products.
Marketing Concept	Focuses on needs/wants of target markets and delivering satisfaction better than competitors. Focuses on needs/wants oftarget markets and delivering superior value
Societal Marketing Concept	Society's well-being. Marketers need to be socially oriented too.

EXCHANGE: Marketing is a Exchange Process

Exchange means:

- ✰ Get something (product/service) by offering something in return. *e.g.* barter or money.
- ✰ Exchange is a value creating process because it leaves normally both parties better off. (win – win situation)

TRANSACTION AND TRANSFER: Marketing involves Transactions

- ✰ A transaction is a trade of values between two or more parties. A transaction is an exchange between **two things of value** on **agreed conditions and a time and place of agreement.** Marketing transactions take place between manufactures and distributors, between distributors and retailers and between retailers and consumers *etc.*

- ☆ A transfer is a one way exchange without receiving anything in return. (gifts, subsidies, charity).

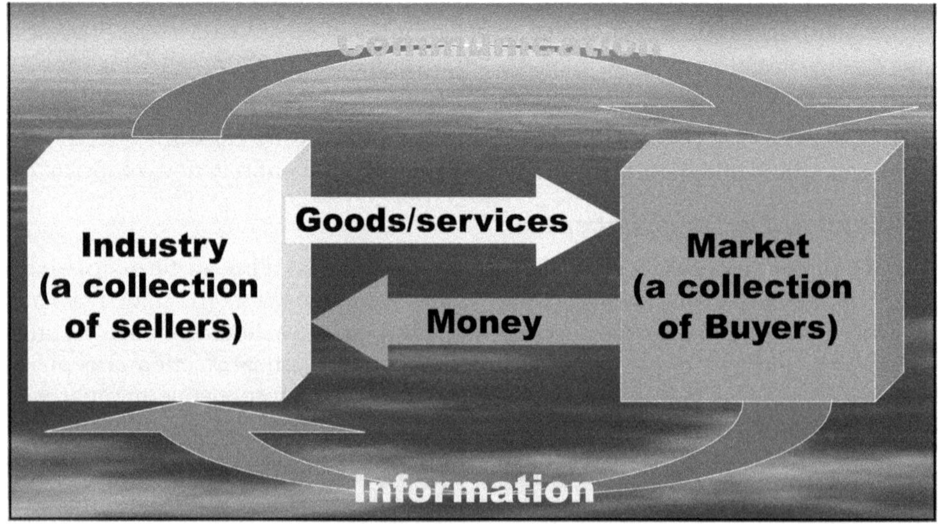

Figure 1.2: Simple Marketing System

In simple words, Industry is collection of sellers, market is collections of buyers and they exchange goods and services for money. Besides it industry or markets communicate their products, prices, availability to buyers/market and give information about their requirements and feedback to companies.

WHAT IS MARKETED? (PRODUCT)

A **PRODUCT** is anything offered to satisfy needs and wants. it includes the following:

- ☆ Goods
- ☆ Services
- ☆ Events
- ☆ Experiences
- ☆ Persons
- ☆ Places
- ☆ Properties
- ☆ Organizations
- ☆ Information
- ☆ Ideas

Marketing people are involved in marketing 10 types of entities: goods, services, experiences, events, persons, places, properties, organizations, information, and ideas.

- **Goods**: Physical goods constitute the bulk of most countries' production and marketing effort. Physical goods, like agricultural produce, equipments, automobiles are marketed.
- **Services**: As economies advance, a growing proportion of their activities are focused on the production of services. Services include banking, insurance, retailing, maintenance and repair, as well as professionals such as accountants, lawyers, consultants, and doctors. Many market offerings consist of a variable mix of goods and services.
- **Experiences**: By orchestrating several services and goods, one can create, stage,and market experiences. Ex. Amusement parks and theme restaurants, agrotourism.
- **Events**: Marketers promote time-based events, such as the exhibition, tradeshows, sports events, and artistic performances.
- **Persons**: Celebrity marketing has become a major business. Artists, sportsman, musicians, CEOs, politicians and other professionals draw help from celebrity marketers.
- **Places**: Cities, states, regions, and nations compete to attract tourists, factories,company headquarters, new residents and investments. Place marketers include economic development specialists, real estate agents, commercial banks, state governments, local business associations, and advertising and public relations agencies.
- **Properties**: Properties are intangible rights of ownership of either real property (real estate) or financial property (stocks and bonds). Properties are bought and sold, and this occasions a marketing effort by real estate agents (for real estate) and investment companies and banks (for securities).
- **Organizations**: Organizations actively work to build a strong, favorable image in the mind of their publics. Universities, companies, museums, and performing arts organizations boost their public images to compete more successfully for audiences and funds.
- **Information**: The production, packaging, and distribution of information is one of society's major industries. Among the marketers of information are schools and universities; publishers of encyclopedias, market, information services, and specialized magazines; makers of CDs; and Internet Web sites.
- **Ideas**: Every market offering has a basic idea at its core. In essence, products and services are platforms for delivering some idea or benefit to satisfy a core need. Example: ideas of polio vaccination, save tigers, organic farming *etc.*

A **BRAND** is when the product is from a known source. An identifying symbol/word/mark distinguishing a product/company from its competitors.

NEEDS, WANTS AND DEMAND

- ✰ Needs are basic human requirements like food, shelter, security *etc.*
- ✰ Wants are needs directed to specific objects/services that might satisfy the need. Wants are many. Wants are shaped by one's society.
- ✰ Demand is the wants for specific products backed by an ability to pay and willingness to buy. Marketers are more interested in knowing the demand.

In this globalised era and competitive world, there is a change from "**make and sell**" philosophy to "**sense and respond**" meaning, marketers should respond to changing customer needs and preferences.

VALUE AND SATISFACTION

The offering will be successful if it delivers value and satisfaction to target buyer.

- ✰ Value = Benefits/Costs
- ✰ Value reflects the perceived tangible and intangible benefits and costs to the customers.
- ✰ Benefits = Functional Benefits + Emotional benefits

  Functional benefits such as performance, economy *etc.*
- ✰ Costs = Monetary costs + Time + Energy + Psychic costs

  Costs include money spent along with time, energy and emotional benefits such as status, joy, satisfaction *etc.*

Value can be seen primarily as a combination of quality, service and psychic costs and price called the customer value triad.

Satisfaction reflects a person's comparative judgments resulting from a product's perceived performance in relation to his/her expectations. It is relative to customer expectations.

CREATION OF UTILITY

Marketing is productive, and is as necessary as the farm production. It is, in fact, a part of production itself, for production is complete only when the product reaches a place in the form and at the time required by the consumers. Marketing adds cost to the product; but, at the same time, it adds utilities to the product. The following four types of utilities of the product are created by marketing:

(a) **Form Utility: The processing function adds form utility to the product by** changing the raw material into a finished form. With this change, the product becomes more useful than it is in the form in which it is produced by the farmer. For example, through processing, oilseeds are converted into oil, sugarcane into sugar, cotton into cloth and wheat into flour and bread.The processed forms are more useful than the original raw materials.

(b) **Place Utility: The transportation function adds place utility to products by** shifting them to a place of need from the place of plenty. Products command higher prices at the place of need than at the place of production because of the increased utility of the product. Ex.: ATM's, home delivery, online purchases.

(c) **Time Utility: The storage function adds time utility to the products by making** them available at the time when they are needed. Ex.: Warehousing, cold storage.

(d) **Possession Utility: The marketing function of buying and selling helps in the** transfer of ownership from one person to another. Products are transferred through marketing to persons having a higher utility from persons having a low utility. Ex.: Retailing *etc.*

DIFFERENCE IN MARKETING OF AGRICULTURAL AND MANUFACTURED GOODS

The marketing of agricultural commodities is different from the marketing of manufactured commodities because of the special characteristics of the agricultural sector(demand and supply) which have a bearing on marketing.

1. Perishability of the Product

Most farm products are perishable in nature; but the period of their perishability varies from a few hours to a few months. Their perishability makes it almost impossible for producers to fix the reserve price for their farm-grown products. The extent of perishability of farm products may be reduced by the processing function; but they cannot be made non-perishable. Nor can their supply be made regular.

2. Seasonality of Production

Farm products are produced in a particular season; they cannot be produced throughout the year. In the harvest season, supply increases and prices fall. But the supply of manufactured products can be adjusted or made uniform throughout the year. Their prices therefore remain almost the same throughout the year, but not so far farm produce.

3. Bulkiness of Products

The characteristic of bulkiness of most farm products makes their transportation and storage difficult and expensive. This fact also restricts the location of production to somewhere near the place of consumption or processing. The price spread in bulky products is higher because of the higher costs of transportation and storage. Processing also needs to be near to production to reduce costs.

4. Variation in Quality of Products

There is a large variation in the quality of agricultural products, which makes their grading and standardization somewhat difficult. There is no such problem in manufactured goods, for they are products of uniform quality. There are large numbers of varieties of some produce with huge quality and price difference.

5. Irregular Supply of Agricultural Products

The supply of agricultural products is uncertain and irregular because of the dependence of agricultural production on natural conditions. With the varying supply, the demand remaining almost constant, the prices of agricultural products fluctuate substantially. This creates problems for farmers, processors and consumers.

6. Small Size of Holdings and Scattered Production

Farm products are produced throughout the length and breadth of the country and most of the producers are of small and marginal. This makes the estimation of supply difficult and creates problems in marketing. This also creates issues of economies of scale and access to markets.

7. Processing

Most of the farm products have to be processed before their consumption by the ultimate consumers. This processing function increases the price spread of agricultural commodities. Processing firms enjoy the advantage of monopsony(only buyer), oligopsony(few buyers) or duopsony(two buyers) in the market. This situation creates disincentives for the producers and may have an adverse effect on production in the next year.

MARKETING MANAGEMENT TASKS

- ☆ Developing marketing strategies and plans
- ☆ Capturing marketing insights : from environment,marketing research.
- ☆ Connecting with customers
- ☆ Building strong brands
- ☆ Shaping the market offerings
- ☆ Communicating value
- ☆ Creating long term growth

An efficient marketing system provides an incentive to farmers to produce more; coveys changing needs of the economy to enable production planning; and fosters competition among traders, and eliminates exploitation, particularly among the small and marginal farmers. The Agriculture market in India today is dominated by rural primary markets that meet local demand, secondary markets that service distant demands and wholesale markets that gather large amounts of produce from different sources for the retailers in the country.

Agricultural marketing has assumed increased importance after launching of the new economic policy and consequent opening up of India's markets to foreign suppliers and buyers and access by Indians to world markets. Efficient marketing practices enable Indian farmers derive the full benefits from the new liberalized world trade regime. A better understanding of Marketing principles in general and Agri-Marketing Practices in particular will remove various constraints and deficiencies in the existing domestic markets and marketing practices.

Chapter 2

Agricultural Marketing Environment

One of the major responsibilities of marketing executives is to monitor and search the environment which is constantly spinning out new opportunities. The marketing environment also spins out new threats such as financial, economic, political and legal and firms find their markets collapsing. Company and marketers need to constantly monitor the changing environment more closely so that they will be able to alter their marketing strategies to meet new challenges and opportunities in the environment.

The marketing environment surrounds and impacts upon the organization. There are three key perspectives on the marketing environment, namely the 'macro-environment,' the 'micro-environment' and the 'internal environment'.

The Micro-Environment

This environment influences the organization directly. It includes suppliers that deal directly or indirectly, consumers and customers, and other local stakeholders. Micro tends to suggest small, but this can be misleading. In this context, micro describes the relationship between firms and the driving forces that control this relationship. It is a more local relationship, and the firm may exercise a degree of influence.

The Macro-Environment

This includes all factors that can influence any organization, but that are out of their direct control. A company does not generally influence any laws (although it is accepted that they could lobby or be part of a trade organization). It is continuously changing, and the company needs to be flexible to adapt. There may be aggressive competition and rivalry in a market. Globalization means that there is always the

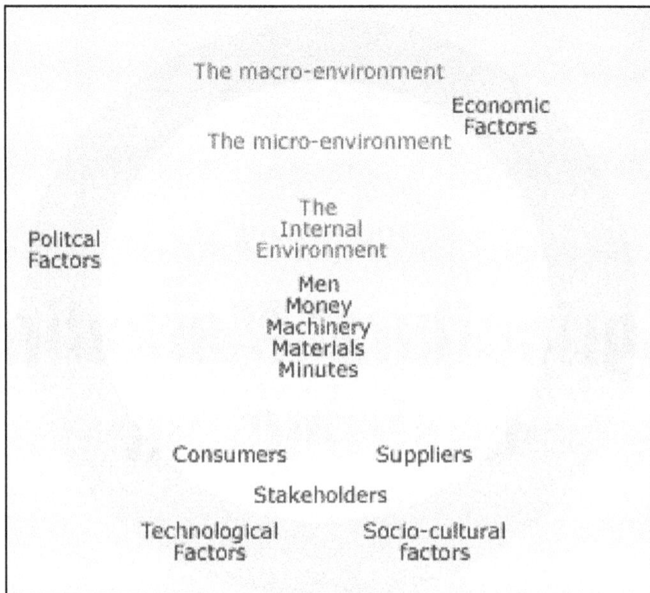

The Marketing Environment.

threat of substitute products and new entrants. The wider environment is also ever changing, and the marketer needs to compensate for changes in culture, politics, economics and technology.

The Internal Environment

All factors that are internal to the organization are known as the 'internal environment'. They are generally audited by applying the 'Five Ms' which are **M**en,Money, **M**achinery, **M**aterials and **M**arkets. The internal environment is as important for managing change as the external. As marketers we call the process of managing internal change 'internal marketing.'

Importance of Environment Analysis

The manager needs to be dynamic to effectively deal with the challenges of environment. The environment of business is not static. Some of the following benefits of environment scanning are as follows:

- It creates an increased general awareness of environmental changes on the part of management
- It guides with greater effectiveness in matters relating to Government
- It helps in marketing analysis
- It suggests improvements in diversification and resource allocation
- It helps firms to identify and capitalize upon opportunities rather than losing out to competitors
- It provides a base of objective qualitative information about the business environment that can subsequently be of value in designing the strategies.

The micro environment consists of the actors in the company's immediate environment that affects its ability to serve the markets: the company, suppliers, market intermediaries, customers, competitors and publics. The macro environment consists of the larger societal forces that affect all of the actors in the company's micro environment the demographic, economic, physical, technological, political, legal and socio-cultural forces.

SUPPLIERS

Suppliers are business firms who provide the needed resource to the company and its competitors to produce the particular goods and services. Any sudden change in the 'suppliers' environment will have a substantial impact on the company's marketing operations. Sometimes some of the inputs to the company might cost more and hence managers have to continuously monitor the fluctuations in the suppliers side. Sudden supply shortage labor strikes and other events can interfere with the fulfilment of delivery promise customers and lose sales in the short run and damage customer goodwill in long run.

MARKETING INTERMEDIARIES

Channel members are the vanguard of the marketing implementation part. They are the people who connect the company with the customers. There are number of middle men who operate in this cycle. Agent middle men like brokers and agents find customers and establish contacts, merchant middlemen are the wholesalers, retailers, who take title to and resell the merchandise. Apart from these channel members, there are physical distribution firms who assist in stocking and moving goods from the original locations to their destinations. Warehouse firms store and protect goods before they move to the next destinations. There are number of transporting firms consists of rail, road, truckers, ship, airline *etc.* that mover goods from one location to another. Every marketer has to decide on the most cost – effective means of transport considering the costs, delivery, safety and speed. There are financial intermediaries like banks, insurance companies, who support the company by providing finance insurance cover *etc.* The behavior and performance of all these intermediaries will affect the marketing operations of the company and the marketing executives have to prudently deal with them.

COMPETITORS

If one company plans a marketing strategy at one side, there are number of other companies in the same industry doing such other calculations. Not only that the competition comes from the branded segment but also from the generic market, where there are only few branded products of rice but there are numerous generic variety of rice according to the local tastes in each region of the country. Basically every company has to identify the competitor, monitor their activities and capture their moves and maintain customer loyalty. Hence every company comes out with their own marketing strategies.

PUBLICS

A public can facilitate or seriously affect the functioning of the company, Philip Kotler defines public as any group that has an actual or potential interest or impact on a company's ability to achieve its objectives. Ex. Customer protection groups, media, environmentalists and governments. Since, the success of the company will be affected by how various publics view their activity, the companies have to monitor these publics, anticipate their moves dealing with them in constructive ways.

CUSTOMERS

Customers are the fulcrum around whom the marketing activities of the organization revolve. The marketer has to face the following types of customer markets.

1. Customer Markets: Markets for personal consumption.
2. Industrial Markets: Goods and services that could become the par tof a product in those industry.
3. Institutional Buyers: Institutions like schools, hospital, which buy in bulk.
4. Reseller Markets: The organizations buy goods for reselling their products.
5. Government Markets: They purchase the products to provide public services.
6. International Markets: Consists of Foreign buyers and Governments.

MACRO ENVIRONMENT

Macro environment consists of six major forces *viz.*, demographic, economic, physical, technological, political/legal and socio-cultural. The trends in each macro environment components and their implications on marketing are discussed below:

Demographic Environment

Demography is the study of human population in terms of size, density, location, age, gender, occupation *etc.* The demographic environment is of major interest to marketers because it involves people the people make up markets. The world population and the Indian population in particular is growing at an explosive rate. This has major implications for business. A growing population means growing human needs. Depending on purchasing powers, it may also mean growing market opportunities. Thus, marketers keep close track of demographic trends developments in their markets and accordingly evolve a suitable marketing programme.

Economic Environment

Markets require purchasing power as well as people. Total purchasing power is functions of current income, prices, savings and credit availability.

Availability of discretionary income shall have the impact on purchasing behavior of the people. Inflation leads consumers to research for opportunities to save money, including buying cheaper brands, economy sizes, *etc.* Consumer expenditures are also affected by consumers savings and debt patterns. The level of savings and borrowings among consumers affect the marketing. Consumption expenditure patters in major goods and services categories have been changing over the years. For instance, when family income rises, the percentage spent on food declines, the percentage spent on housing and house hold operations remain constant, and the percentage spent on other categories such as transportation and education increase. These changing consumer expenditure patterns has an impact on marketing and the marketing executives need to know such changes in economic environment for their marketing decisions.

Natural Environment

There are certain finite renewable resources such as wood and other forest materials which are now dearth in certain parts of world. Similarly there are finite non-renewable resources like oil coal and various minerals, which are also short in supply. In such cases, the marketers have to find out some alternative resources. For instance, the marketers of wooden chairs, due to shortage and high cost of wood shifted to steel and later on fiber chairs. Similarly scientists all over the world are constantly trying to find out alternative sources of energy for oil due to dearth in supply. There has been increase in the pollution levels in the country due to certain chemicals. In metros people are facing increased pollution due to the presence of different industries. Marketers should be aware of the threats and opportunities associated with the natural environment and have to find alternative sources of physical resources.

Socio-Cultural Environment

The socio-cultural environment comprises of the basic beliefs, values and norms which shapes the people. People in a given society hold many core beliefs and values, that will tend to persist. People's secondary beliefs and values are more open to change. Marketers have more chances of changing secondary values but little chance of changing core values. Each society contains sub-cultures, *i.e.* groups of people with shared value systems emerging out of their common life experiences, beliefs, preferences and behaviors. To the extent that sub-cultural groups exhibit different wants and consumption behavior, marketers can choose sub-cultures as their target markets. Secondary cultural values undergo changes over time. Marketers have a keen interest in anticipating cultural shifts in order to identify new marketing opportunities and threats.

Technological Environment

Technology advancement has benefited the society and also caused damages. Technology is accelerating at a pace the many products seen yester-years have become obsolete now. There could be a new range of products and systems due to the innovations in technology. This technology developments has tremendous impact on marketing and unless the marketing manager cope up with this development it cannot survive in the competitive market.

Political and Legal Environment

Marketing decisions are highly affected by changes in the political/legal environment. The environment is made up of laws and government agencies that influence and constraint various organizations and individuals in society. Legislations affecting business has steadily increased over the years.

The legal enactments and rules and regulations exercise a specific impact on the marketing practices, systems and institutions in the country. Some of the acts which have direct bearing on the marketing of the company include, the Prevention of Food Adulteration Act (1954), The Drugs and Cosmetics Act (1940), The Standard Weights and Measures Act (1956) *etc.* Similarly, when the government changes, the policy relating to commerce, trade, economy and finance also changes resulting in changes in business. Hence, the marketing executives needs a good working knowledge of the major laws affecting business and have to adapt themselves to changing legal and political decisions.

All the above micro environmental actors and macro environmental forces affect the marketing systems individually and collectively. The marketing executives need to understand the opportunities and threats caused by these forces and accordingly they must be able to evolve appropriate marketing strategies.

CHANGING AGRICULTURAL MARKETING ENVIRONMENT

Agriculture has evolved into agribusiness and has become a vast and complex system that reaches for beyond the farm to include all those who are involved in bringing food and fiber to consumers. Agribusiness include not only farmers that farm the land but also the people and firms that provide the inputs (for ex. Seed, chemicals, credit *etc.*), process the output (for ex. Milk, grain, meat *etc.*), manufacture the food products (for ex. ice cream, bread, breakfast cereals *etc.*), and transport and sell the food products to consumers (for ex. restaurants, supermarkets).

Agribusiness system has undergone a rapid transformation as new industries have evolved and traditional farming operations have grown larger and more specialized. The transformation did not happen overnight, but came slowly as a response to a variety of forces.

Initially agriculture being the major venture it was easy to become a farmer, but productivity was low. Average farmer produced enough food to feed just four people. As a consequence most farmers were nearly totally self-sufficient. They produced most of the inputs they needed for production, such as seed, feed and simple farm equipment. Farm families processed the commodities they grew to make their own food and clothing. They consumed or used just about everything they produced. The small amount of output not consumed on the farm was sold for cash. These items were used to feed and cloth the minor portion of the country's population that lived in villages and cities. A few agricultural products made their way into the export market and were sold to buyers in other countries.

Farmers found it increasingly profitable to concentrate on production and began to purchase inputs they formerly made themselves. This trend enabled others to build business that focused on meeting the need for inputs used in production of agriculture such as seed, fencing, machinery and so on. These farms involved into the industries that make up the "agricultural inputs sector". Input farms are major part of agribusiness and produce variety of technologically based products that account for approximately 75 per cent of all the inputs used in agriculture.

At the same time the agriculture input sector was evolving, a similar evolution was taking place a commodity processing and food manufacturing moved off the farm. The form of most commodities (wheat, rice, milk, livestock and so on) must be changed to make them more useful and convenient for consumers. For ex. consumers would rather buy flour than grind the wheat themselves before baking a cake. They are willing to pay extra for the convenience of buying the processed commodity (flour) instead of the raw agriculture commodity (wheat).

During the same period technological advances were being made in food preservation method. Up until this time the perishable nature of most agriculture commodities meant that they were available only at harvest. Advance in food processing have made it possible to get those commodities all throughout the year. Today even farm families use purchased food and fiber products rather than doing the processing themselves. The farms that meet the consumers demand for greater processing and convenience also constitute a major part of agribusiness and are referred to as the processing sector.

It is apparent that the definition of agriculture had to be expanded to include more than production. Farmers rely on the input industries to provide the products and service they need to produce agricultural commodities. They also rely on commodity processors, food manufactures, and ultimately food distributors and retailers to purchase their raw agricultural commodities and to process and deliver them to the consumer for final sale. The result is the food and fiber system.

There are five major trends that are affecting agribusiness industry and agriculture. These are

- ☆ Globalization and integrated markets: By 2050 the world food demand is expected to double. Asia is the major source of demand increase because of population and income increase.
- ☆ Information technology: Advancement in IT creates opportunities in e-retailing and ICT enabled supply chain.
- ☆ Biotechnology: It has opened opportunities in Tissue culture, Biopesticides and various other applications.
- ☆ Expanding consumer power: Consumers have become more empowered, aware and quality and health conscious.
- ☆ Mergers/consolidation/strategic alliances among agribusinesses.

With the dramatic changes occurring in the agricultural industries, it is critical for both firms and individuals to develop and maintain competencies that will enhance a competitive position in this rapidly evolving market. The skills or capacities required to be successful are dynamic capabilities which embrace new ideas, change, innovation, analysis, and integration, and teamwork— proficiencies perhaps not part of the experience base in the more traditional agriculture of the past. Developing these capacities will enable participants to be successful players in the new agriculture of integrated, interdependent value chains competing on cost, quality, and time metrics.

Chapter 3

Market Segmentation, Targeting and Positioning

The purpose for segmenting a market is to allow your marketing/sales program to focus on the subset of prospects that are "most likely" to purchase your offering. If done properly, this will help to insure the highest return for your marketing/sales expenditures. Depending on whether you are selling your offering to individual consumers or a business, there are definite differences in what you will consider when defining market segments.

Market Segmentation

It is the sub-dividing of market into homogenous sub-sections of customers, where any sub-section may conceivably be selected as a target market to be reached with distinct marketing aspects.

"Market segmentation is the process in marketing of dividing a market into distinct subsets (segments) that behave in the same way or have similar needs."

Because each segment is fairly homogeneous in their needs and attitudes, they are likely to respond similarly to a given marketing strategy. That is, they are likely to have similar feelings and ideas about a marketing mix comprised of a given product or service, sold at a given price, distributed in a certain way and promoted in a certain way.

Bases for Market Segmentation

Broadly, markets can be divided according to a number of general criteria, such as by industry or public versus private sector. Small segments are often termed niche markets or specialty markets.

However, all segments fall into either consumer or industrial markets.

The common bases of marketing segmentation used for consumer markets are as follows:

- ☆ Geographic segmentation (climatic conditions, states, type of city, International, urban/rural)
- ☆ Demographic segmentation (Income, Age, gender, education, ethnicity, and family life cycle stage, family size)
- ☆ Psychographic segmentation (Personality, Lifestyle and Values)
- ☆ Behaviourial segmentation (Usage rate, occasions, attitude, benefit seeked, Loyalty status *etc.*)

Segmentation of Markets for Industrial Goods

An industrial market is vastly different from a consumer goods market. It can be segmented on the basis of:

1. Customer size
2. Purchasing methods and policies
3. Geographical location
4. Kind of organization

The following classification of industries provides usually the basis for segmenting an industrial market.

1. Services market
2. Agriculture, forestry and fishing
3. Manufacturing industries
4. Transportation, communications and other public utilities
5. Construction
6. Mining and quarrying

The process of segmentation is distinct from targeting (choosing which segments to address)and positioning (designing an appropriate marketing mix for each segment). The overall intent is to identify groups of similar customers and potential customers; to prioritize the groups to address; to understand their behavior; and to respond with appropriate marketing strategies that satisfy the different preferences of each chosen segment. Revenues are thus improved.

Improved segmentation can lead to significantly improved marketing effectiveness. With the right segmentation, the right products can be designed, advertising results can be improved and customer satisfaction can be increased.

Segmentation, targeting, and positioning together comprise a three stage process. We first (1) determine which kinds of customers exist, then (2) select which ones we are best off trying to serve and, finally, (3) implement our segmentation by optimizing our products/services for that segment and communicating that we have made the choice to distinguish ourselves that way

Segmentation, Targeting, Positioning

- ***Segmentation***: grouping consumers by some criteria
- ***Targeting***: choosing which group(s) to sell to
- ***Positioning***: select the marketing mix most appropriate for the target segment(s)

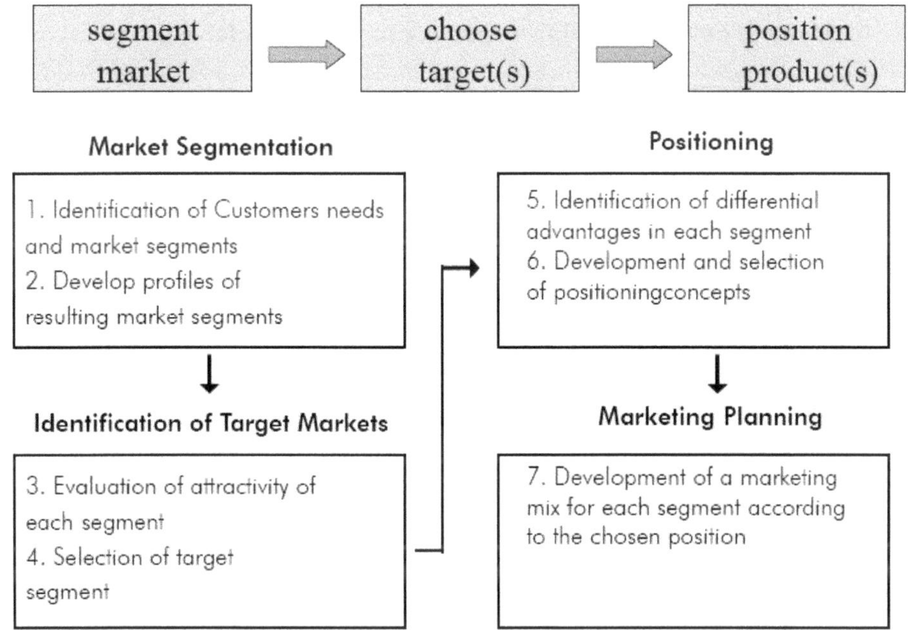

Our choice of segmentation should generally depend on several factors.

First, how well are existing segments served by other manufacturers?

Secondly, how large is the segment, and how can we expect it to grow?

Thirdly, Do we have strengths as a company that will help us appeal particularly to one group of Consumers?

Market Targeting

It is the act of evaluating and comparing the identified groups and then selecting one or more of them as the prospects with the highest potential.

Selecting a Target Market Strategy

There are three basic target market strategies

1) Undifferentiated Marketing

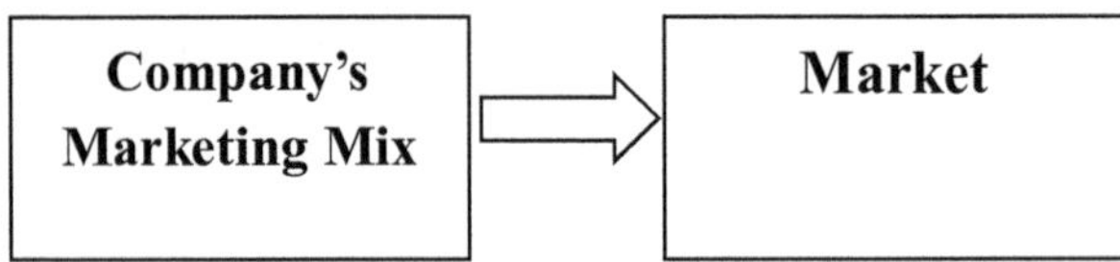

It is a marketing strategy created by the same marketing mix product, price, place and promotion for all markets. The firm requires extensive distribution in the maximum number of retail outlets, *e.g.* most of agricultural commodities sold undifferentiated.

2) Differentiated Marketing

It involves devising multiple marketing mix offerings for different segments.

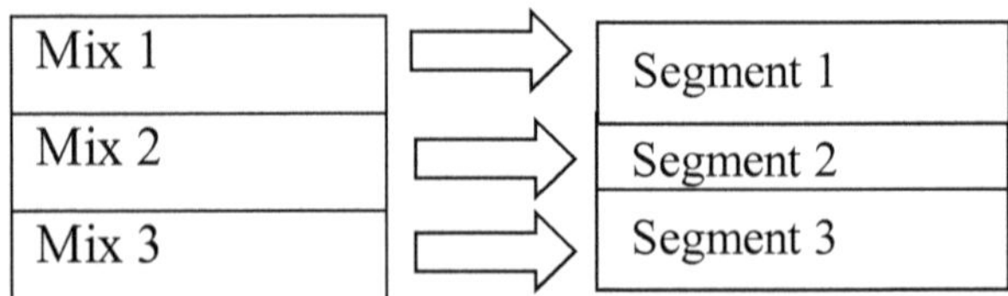

for *e.g.* Milk with different fat contents for different segments.

3) Concentrated Marketing

It involves devising a marketing mix on one product, one segment principle. For example, Johnson and Johnson has chosen to position itself as a maker of baby care products, Tractor manufacturer focussing only on Agricultural tractor market.

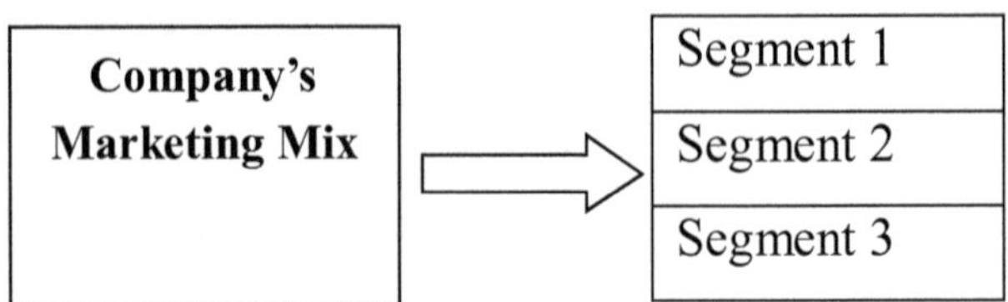

Hence, it can be seen that targeted marketing consists of segmenting the market, choosing which segments to serve and designing the marketing mix in such a way that it is attractive to the chosen segments.

Positioning

Positioning involves implementing our targeting. Positioning is the placement of a product, service, outlet, *etc.* in the mind of the consumer. *Al Ries and Jack Trout* coined this term. They see positioning as a creative exercise done with a product. It includes:

- Determine what consumers currently think about your product (with respect to competing products)
- Decide what you *want* consumers to think about your product.
- Designing the company's offering and image to occupy a distinctive place in the mind of the target market.
- Result of positioning strategy is successful creation of a *customer–focused value proposition.*
- Positioning is not what you do to the product but what you do the mind of the prospect.

Positioning takes into account the uniqueness of a company's marketing mix in relation to that of competitors. The uniqueness or differentiation may be tangible or intangible depending upon the physical attributes or the psychological attributes of the product. Establishing and communicating these distinctive aspects is termed as positioning, *e.g.*, Kohinoor Foods positioned its Basmati rice in UK market as "True Basmati" for authentic Indian Basmati rice.

Repositioning involves an attempt to change consumer perceptions of a brand, usually because the existing position that the brand holds has become less attractive.

Chapter 4

Consumer Behaviour

Buyer/consumer behaviour may be defined as the activities and decision processes involved in choosing between alternatives, procuring and using products or services. The truth is that marketing may promote a given product, service or practice but unless the target audience perceives that product, service or practice to be relevant to their needs then they will never try it. Moreover, unless their first time trial of the product, service or practice is positive, they will not try it a second time. The purpose of studying buyer behaviour is to better meet the needs of customers. Only by doing so will the marketing enterprise continually and consistently meet its own needs.

Factors Affecting Buyer Behaviour

The behaviour of buyers is the product of two broad categories of influence; these are endogenous/internal factors (*i.e.* those internal to the individual) and exogenous/external factors (*i.e.* those external to the individual). The most important of these two categories of factors are depicted in Figure 4.1.

Exogenous Factors Affecting Buyer Behaviour

Factors which are external to the individual but have a substantial impact upon his/her behaviour are social and cultural in nature. These include culture, social class or status, reference groups and family membership.

a) Culture

Culture is perhaps the most fundamental and most pervasive external influence on an individual's behaviour, including his/her buying behaviour. Culture has been defined as:

"...the complex of values, ideas, attitudes and other meaningful symbols created by people to shape human behaviour and the artifacts of that behaviour as they are transmitted from one generation to the next."[1]

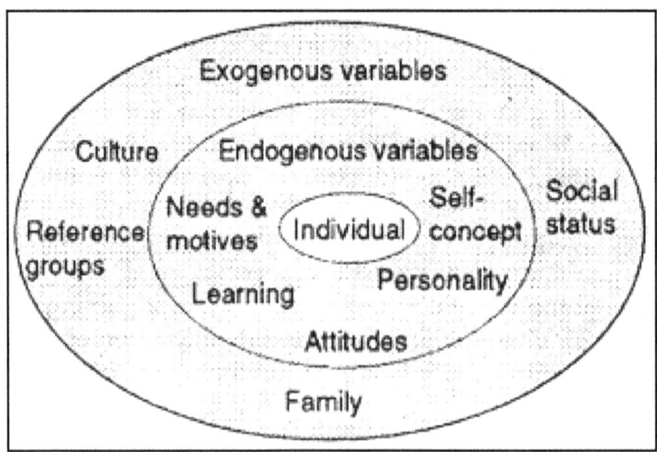

Figure 4.1: Endogenous and Exogenous Factors Impinging upon Buyer Behaviour.

Culture is the mechanism by which each society evolves its distinctive behavioural patterns and values and transmits these to subsequent generations. Without a knowledge of the culture into which a product is being marketed mistakes can be made and opportunities missed. Creative marketers who do have a knowledge of cultural norms and values can profit by aligning product benefits and characteristics with these social standards.

Within any particular society the culture will comprise of a number of subcultures. That is, there will be various racial, ethnic and religious groups. Each, to some degree, will have distinct beliefs and values. Food is very culturally sensitive item. Different food products are consumed during special occasions and festivals. India has different states which have different food habits associated with their culture.

b) Social Status

Social class or social status is a powerful tool for segmenting markets. Research suggests that people from the same social group tend to have similar opportunities, live in similar types of housing, in the same areas, buy similar products from the same types of outlets and generally conform to similar styles of living. The variables used to stratify a population into social classes or groups normally include income, occupation, education and lifestyle. Every status has its roles - a set of proper behaviors specified by culturally defined rules. A group influences its members primarily through the roles and behavioral norms expected of them.

Thus, the behaviour of an individual, on a given occasion, will relate to the social role which he/she is acting out. The marketer needs to know what role a person of a given status is playing and what is expected of that individual by the group which has conferred the status upon him/her. Such an understanding can significantly affect the marketing strategy employed with respect to that category of customer. Agri input companies and extension organisations give demonstrations of improved inputs and technologies at the identified progressive farmers field or community leaders as they have significant role and influence on other farmers.

c) Reference Groups

People are social animals who tend to live in groups. The group(s) to which a person belongs exerts an influence upon the behaviour, beliefs and attitudes of its members by communicating norms and expectations about the roles they are to assume. Thus, an individual will refer to others with respect to: 'correct' modes of dress and speech; the legitimacy of values, beliefs and attitudes; the appropriateness of certain forms of behaviour, and also on the social acceptability of the consumption of given products and services. These "others' constitute reference groups. Reference groups provide a standard of comparison against which an individual can judge his/her own attitudes, beliefs and behaviour.

An individual need not belong to a given group in order for that group to exert an influence upon his/her behaviour. *Shibutani* has identified three distinct reference groups:

- A group to which an individual belongs (also known as a peer group)
- A group to which an individual aspires, and
- A group whose perspective has been adopted by the individual

A small farmer will identify with other farmers whose farm and operations are similar in size and technology and will feel that he/she belongs to this group. He/she may have ambitions to become a progressive farmer adopting modern technology and so aspires to membership of a group recognised as that of progressive farmers. At the same time, the small farmer may adopt the views and opinions of a Farmers association since he/she believes that when this group voices an opinion about trends or proposed changes to the trade their arguments are normally in the best interests of small farmers. The common factor between these three groups is that they each provide a frame of reference for the individual. Reference groups can have a significant influence on patterns of product use and consumption also.

d) Families as Reference Groups

The family is another group which influences the behaviour of individuals including buying behaviour. Two types of family may be distinguished from one another, the nuclear family and the extended family. The nuclear family is the basic family unit and describes the parents and immediate off-spring and/or their adopted children. The extended family includes all living relatives in addition to the parents and their children - grandparents, aunts, uncles, cousins, step-relatives and in-laws (*i.e.* relatives through marriage).

Families often form a Decision-Making Unit (DMU) with respect to household purchases, with each member performing a different role. For instance, the children may initiate the purchase by requesting a breakfast cereal in place of bread, the male head of the household may decide whether a certain category of purchase may be made such as this more expensive type of breakfast food and the female head of the household may contribute to the decision to buy a processed breakfast food and decide which brand and from which retail outlet it is to be bought. Where the extended family becomes involved in a purchasing decision the DMU becomes larger and the roles of family members more diverse. When marketing to families it is essential to know which members play a role in certain types of decision and what role they play. Thus, for instance, the cereals manufacturer may target mass media advertising at children since they trigger a purchase whilst in-store merchandising and promotion is designed to appeal to the housewives or other female heads of household because they make the brand choice.

Endogenous Influences on Buyer Behaviour

Endogenous influences are those which are internal to the individual. These are psychological in nature and include needs and motives, perceptions, learning processes, attitudes, personality type and self-image.

a) Needs and Motives

The terms needs and motivations are often viewed to be interchangeable. However, there is a difference between them. When an individual recognises that he/she has a need, this acts to trigger a motivated state. Need recognition occurs when the individual becomes aware of a discrepancy between his/her actual state and some perceived desired state. The housewife who buys polished dal (pulse) (actual state), who is made aware of the vitamin deficiencies in these products and is anxious to be, and to be seen to be, a wife and/or mother who looks after the health of her family (desired state) could be motivated to purchase less refined or unpolished dal (pulse). More formally, a need is a perceived difference between an ideal state and some desired state which is sufficiently large and important to stimulate a behavioural reaction. Once the need is recognised then the individual concerned will form a motive. A motive may be defined as an impulse to act in such a way as to bring about the meeting of a specific need.

b) Perceptions

Whereas motivation is a stimulus to action, how an individual perceives situations, products, promotional messages, and even the source of such messages, largely determines how an individual acts. A basic definition of perception would be 'how people see things'. Perceptionis defined as the process by which an individual selects, organizes, and interprets information inputs to create a meaningful picture of the world.

Individuals can have vastly differing interpretations of the same situation. Moreover, individuals can hold widely differing perceptions, or interpretations, of the same stimulus due to three perceptual processes, *i.e.* selective attention, selective distortion and selective retention.

Selective Attention

All people are daily bombarded by stimuli, both commercial and non-commercial. People simply cannot pay attention to all these messages and therefore they develop mechanisms to reduce the amount of information that they actually process. People pay attention to stimuli which meet an immediate need.

Selective Distortion

Incoming information is often distorted to fit existing beliefs, opinions and expectations. Such beliefs are based on perceptions rather than experiences.

Selective Retention

People forget all too easily. The information retained is generally that which supports the decision maker's existing attitudes and beliefs. Thus a consumer who is strongly loyal to a particular brand of Basmati Rice will easily recall the benefits claimed for that product in advertising campaigns but will forget the claims made for a competing product.

c) Learning

Much of human behaviour is learned. The evidence of learning is a change in a person's behaviour as a result of experience. Theory suggests that learning is the product of interactions between drives, stimuli, cues, responses and reinforcement. For instance, a farmer may have a strong drive towards increasing his/her productivity. A drive is a strong internal stimulus impelling action. Drive turns to motive when it focuses upon a particular drive-reducing stimulus object. A farmer may see the adoption of a newly available tractor as a way of increasing his/her productivity to the extent required. The farmer's response to the notion of buying a tractor is influenced by the surrounding cues. A cue is a lesser stimulus that can determine whether an individual responds and, if so, how he/she responds. The encouragement of the farmer's neighbours, and perhaps his/her village headman, seeing the same type of tractor operating successfully on a neighbouring farm, receiving visits from salesmen and reading promotion literature are all cues that can impinge upon the farmer's impulse to invest in a tractor.

If the farmer buys the tractor and if he/she finds that it works well and improves his/her productivity to the level required, then learning is positively reinforced. If the buyer's experience does not match expectations then he/she is likely to suffer cognitive dissonance. Cognitive consistency theories hold that individuals strive to maintain a consistent set of attitudes and beliefs. When attitudes and beliefs about a product or service are challenged, due to its performance falling short of expectations, then the buyer experiences an uncomfortable psychological state and becomes motivated to redress the balance between expectations and experience.

Market oriented organisations have policies which seek to deal with cognitive dissonance. No matter how much care an organisations takes in the manufacture and distribution of its products it is unlikely to achieve 'zero defects' all of the time. Consequently, some buyers will be dissatisfied at some point in time. The fact that this happens is less important than how the company deals with dissatisfaction. Many companies operate a policy of giving buyers a choice of having their money back or accepting a replacement of product.

d) Attitudes

Fishbein and Ajzen put forward a definition of attitudes which has become widely accepted. Their definition is:

"...a learned predisposition to respond in a consistently favourable or unfavourable manner with respect to a given object."

This definition draws attention to four fundamental characteristics of attitudes. First it suggests that attitudes are enduring. They may change over time but they tend to be reasonably stable in the short to medium term. Second, the definition stresses that attitudes are learned from the individual's own experience and/or from what they read or hear from others. Third, that attitudes precede and impact upon behaviour. Attitudes reflect an individual's predispositions towards another person, an event, product or other object. A person may be either favourably or unfavourably predisposed towards an object; or they may be indifferent towards that object and therefore fail to display any behavioural pattern with respect to the object. Fourth, the chief function of attitudes is to facilitate the evaluation of objects. Attitudes are a generalisation and therefore the individual does not have to go through a process of evaluation tailored to each and every object.

Marketers have to work hard at creating positive attitudes towards the organisation, its products or services and any intermediaries it may channel these products/services through. Changing negative attitudes requires even more effort. It is generally more difficult, and expensive, to change a negative attitude than to cultivate a positive attitude at the outset. Indeed, it is usually more productive to make changes to the product's characteristics and/or image, to fit the existing attitudes of buyers, than to seek to change firmly entrenched attitudes.

e) Personality and Self-Concept

Individuals tend to perceive other human beings as 'types of persons'. There are, for example, people perceived to be nervous types, ambitious types, self-confident types, introverts, extroverts, the timid, the bold and so on. These are personality traits. Like attitudes, personality traits serve to bring about a consistency in the behaviour of an individual with respect to his/her environment. Thus, for example, a personality characterised by a high degree of self-confidence will consistently be outspoken with respect to his/her views on new ideas, products, processes and practices. Moreover, where there is an element of risk in adopting an innovative product the self-confident personality will be more often among the risk-takers than the risk-averse.

The Consumer Buying Decision Process

Buying decisions may be made by individuals or a group such as a family or a committee within a commercial or industrial organisation. Where a group is involved, the term Decision-Making Unit (DMU) is commonly used. Marketers are interested in identifying all of the parties involved in the decision making process and are careful to distinguish between buyers and users. The farmer may make the final decision as to whether a given piece of agricultural equipment is purchased but his/her decision could well be influenced by the views, attitudes and amptitudes of the farm worker who will operate the machine. Moreover, the subsequent experience of the operator will play a major role in determining whether or not the decision to buy is positively reinforced. Similarly, the mother in the family may be the chief buyer of household foods but children may have a major influence on the purchase of those food items of which they are the main consumers.

Typically, the buying decision models comprised five stages as shown in figure below: problem recognition, information search, evaluation of alternatives, purchase decision and post-purchase behaviour. Such models underline the fact that the actual decision to purchase is but a single event in a process which begins sometime beforehand and continues after the item is bought. The marketer is encouraged to think about influencing a buying process rather than a buying decision.

Figure 4.2: A Five-Stage Model of the Buying Process.

Problem Recognition

The buying process begins with a recognition on the part of an individual or organisation that they have a problem or need. The farmer recognises that he/she is approaching a new cultivation season and requires seed; a grain trading company realises that stocks are depleted but demand is rising and therefore wheat, rice and maize must be procured.

Problems and needs can be triggered by either internal or external stimuli. A farmer may purchase based on his own experience (internal stimuli) or based on advice by Retailer (external stimuli).

Marketing research needs to identify the stimuli that trigger the recognition of particular problems and needs. Research should be directed towards establishing the needs/problems that arose, how these were brought about and how buyers arrived at the decision that a particular product was likely to meet their need or solve their problem. By so doing marketers can design products/services capable of meeting those needs/problems and develop marketing strategies that can trigger customer interest in those products or services.

Information Search

Information gathering may be passive or active. Passive information gathering occurs when an individual or group simply becomes more attentive to a recognised solution to a given need. That is, he/she exhibits heightened attention. The potential buyer becomes more aware of advertisements or other messages concerning the product in question. In other circumstances the individual is proactive rather than reactive with respect to information. A trader who sees potential in a new vegetable which is being imported into the country will actively search out information about the product, sources of supply, prices and import regulations. He/she is likely to converse with other traders, request literature from potential suppliers, *etc.*

Marketers will be interested to establish what information sources tend to seek out. Kotler[8] states that the information sources used will fall into four categories:

- Personal sources (family, friends, work colleagues, neighbours, acquaintances)
- Commercial sources (promotional materials, press releases, technical journals or consumer magazines, distributors, packaging)
- Public sources (mass media)
- Experiential (handling, using the product).

Kotler suggests that, in the case of consumers, these sources of information will play different roles. It is generally held that communications from commercial and other non-personal sources provide information whilst personal sources, such as family or friends, help in evaluating a product or in making choices between alternatives. The extent of information seeking will vary with the intensity of the drive to 'solve' the problem and the amount of information that the individual already possesses.

As an individual engages in information gathering he/she becomes more knowledgeable about the range of alternative products or brands available. In highly competitive markets where there is a large number of competing products or brands the customer rarely makes a choice from the entire set of alternatives available. Rather, the customer selects from a subset of the alternative products or brands that are actually available, termed 'the evoked set'. Figure 4.3 illustrates the process involved in arriving at the evoked set, *i.e.* the set of products/brands from which the customer actually chooses.

Since a customer's information is likely to be imperfect he/she will be ignorant of the existence of a number of products/brands that are actually available on the market. This happens for a variety of reasons. The customer may only engage in limited information gathering, some products/brands may not be strongly promoted or some may be heavily promoted in distribution channels that a particular customer does not frequent. Thus, the customer is seldom in a position to choose products/ brands from the total set. Rather, the customer is only aware of a subset of the total set. Some of these will fail to meet the customer's initial screening criteria. Some will lie outside the customer's price range (they may be either too expensive or too cheap), some will have too high or too low a specification, others might not have

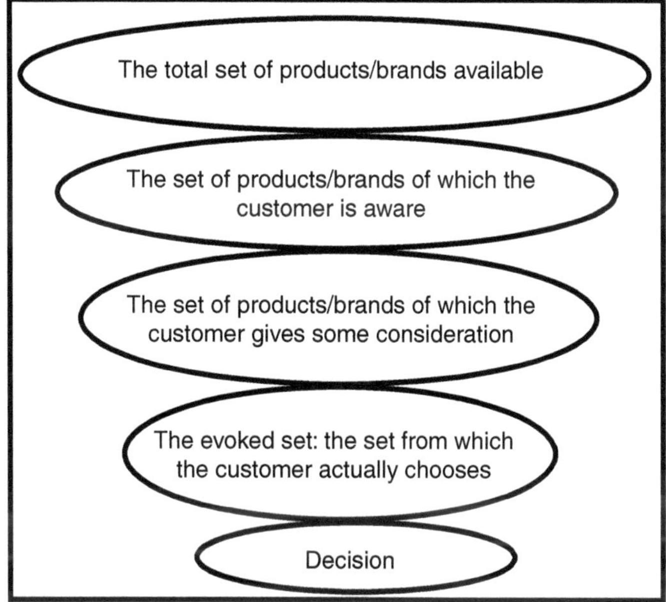

Figure 4.3: The Concept of an Evoked Set.

the basic level of technical service support. Therefore, the set of products/brands of which the customer is aware is then reduced to a further subset of products/brands to which the customer gives serious consideration. However, as the prospective customer gathers more information the set of alternatives is further reduced until he/she arrives at an evoked set. This is the set of alternative products or brands from which a customer's actual choice is made.

The important implication of the evoked set theory for marketing managers is that they must know when their products are failing to get into the evoked set and should determine what criteria potential customers are using as a basis for including and excluding products/brands from their evoked sets. It is equally important, although not always easy, to establish what information sources customers are using and the roles and relative importance of alternative sources.

Evaluation of Alternatives

The process of evaluating alternatives not only differs from customer to customer prospective customer but the individual will also adopt different processes in accordance with the situation. It is likely that when making judgments customers will focus on those product attributes and features that are most relevant to their needs at a given point in time. Here, the marketer can differentiate between those characteristics which a product must have before it is allowed to enter the customer's evoked set. A quite different set of criteria might be used in deciding between alternative products and suppliers within the evoked set *e.g.* the period of credit given by the supplier, the ability of the supplier to deliver the total order in periodic batches and the reliability of the supplier in the past.

Purchase Decision

At the evaluation stage the prospective customer will have arrived at a judgement about his/her preference among the evoked set and have formed a purchase intention. However, two factors can intervene between the intention and the purchase decision: the attitude of others and unanticipated events. If the attitude of other individuals or organisations who influence the prospective customer is strongly negative then the intention may not be converted to a firm commitment or decision. Unanticipated events can also intervene between intention and action. Whenever human beings form judgements or seek to make decisions they invariably make assumptions. These assumptions are often implicit rather than explicit. A farmer may state an intention to purchase a mechanical thresher within the next twelve months but when his/her implicit assumption of 'a good harvest' is not realised, due to drought, the purchase of the machine is postponed.

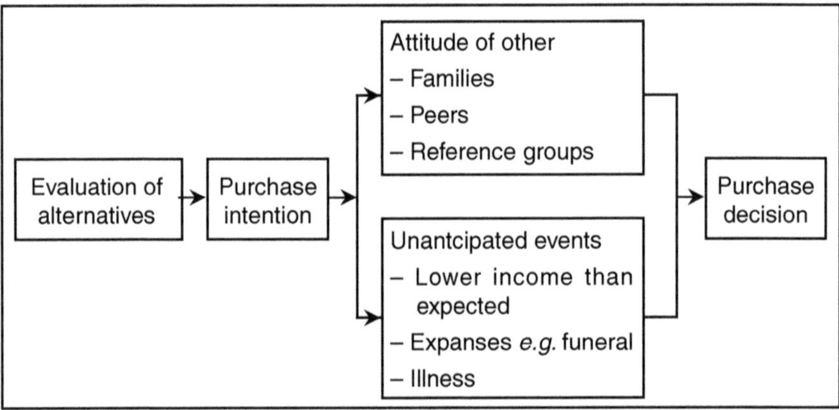

Figure 4.4: Factors Intervening between Intention and Purchase

Postpurchase Behaviour

The process of marketing is not concluded when a sale is made. Marketing continues into the postpurchase period. The aim of marketing is not to make a sale but to create a long term relationship with a customer. Organisations maintain profitability and growth through repeat purchases of their products and services by loyal customers.

Having procured the product the customer will experience either satisfaction or dissatisfaction with his/her purchase. The level of satisfaction or dissatisfaction is largely a function of the congruence between the buyer's expectations of the product and the product's perceived performance. Buyer expectations of a product are usually based upon promotional messages from the product's supplier, family, friends, work colleagues and, perhaps, professional advisors. In addition, the buyer's own perceptual processes influence expectations. If the product's perceived performance either matches or exceeds its expected performance then the buyer is likely to feel highly satisfied. It is in the best long term interests of commercial organisations not to oversell their products. That is, the claims made for products should faithfully reflect the product's actual performance capabilities. Even then, this will not prevent some buyers from holding unreasonable expectations of the product.

Another aspect of postpurchase behaviour that is of interest to marketers is how the buyer actually uses the product. It is common to find buyers using a product in a different way from that for which it was either designed or intended. Such deviations can present problems or opportunities to the product supplier.

Buyers do not invariably pass through all five stages described here. Much depends upon the circumstances surrounding the purchase decision. In the case of less expensive and/or frequently purchased items there would probably be far less searching for information. If the prospective buyer is loyal to a particular brand then the evaluation of alternatives may not figure at all. The fact that some of the stages depicted may be skipped, in certain circumstances, does not invalidate the model. The five-stage model outlined here shows the complete sequence of possible events in the buying process.

Organisational Buying Behaviour

Organisational buying behaviour has been defined by Webster and Wind as

"...the decision-making process by which formal organizations establish the need for purchased products and services, and identify, evaluate, and choose among alternative brands and suppliers."

Organisational buying decisions are likely to be made by a group rather than wholly by an individual. Webster and Wind coined the term 'decision making unit' (DMU) to describe such groups. Members of DMUs may have different roles, including:

Users

The individuals who are most likely to use the product often initiate the buying process by signalling a need for it and outlining its specifications, *e.g.* the production manager in a dairy factory might identify the need to buy a certain amount of milk for yoghurt production and will specify the butter fat content and total bacterial count acceptable in the milk.

Influencers

Others in an organisation may have an influence on the specifications and also provide information on alternatives, *e.g.* the quality control personnel in the dairy factory could amend the production manager's specification to reflect a range of acceptable butter fat content and provide an assessment of the prospects of substituting spray dried milk for fresh milk.

Decision-makers

Decision-makers ultimately have the power to reach conclusions on the product, product specifications and/or suppliers.

Approvers

There could be some person with authority to sanction the purchase specified by the decision-maker.

Buyers

Buyers have the formal responsibility for choosing suppliers and agreeing the terms and conditions attached to the contract of sale.

Gatekeepers

Access to members of the DMU may be controlled by secretaries, switchboard operators, personal assistants and the like. These individuals may also filter information intended for members of the DMU.

When attempting to market to organisations it is imperative that information be gathered on the structure of the DMU within each target enterprise. Intelligence gathering should include attempts to understand the relative roles of each identified member of the DMU. It is also important to appreciate the status or authority of members of the DMU. For example, in some instances the term 'buyer' is used to describe individuals who simply perform the clerical duty of executing the buying decisions made by decision-makers, but in others the 'buyer' is the decision-maker. Moreover, in some organisations the 'buyer' is a member of middle management whereas in others he/she is a member of senior management. This can have implications for the structure and titles of the salesforce in a marketing organisation. Senior managers may be happy to deal with a Sales Manager but not a salesman. It is not simply the sales-person's title which might matter but the extent to which he/she is empowered to make decisions without reference back to a more senior person. Thus, when marketing to organisations it is important that the sales-person is able to deal with the professional buyer on an equal footing, *i.e.* similar status within their respective organisations, similar levels of responsibility and authority and, possibly, job titles that convey their equality.

Organisational Markets

Organisational markets fall into one of three categories: industrial markets, reseller markets and government markets.

Industrial Markets

Industrial buyers procure raw materials, components, semi-finished goods and services as input to the production of other goods and services. Thus industrial markets are characterised by derived demand. That is, the demand for industrial goods ultimately derives from the demand for consumer goods. Since this is the case marketers of industrial products must maintain an interest in patterns of consumer demand and the forces which shape that demand.

Most large industrial commercial organisations employ professionally trained buyers. The tasks of industrial buyers can be categorised as straight rebuys, modified rebuys and new tasks. Some of these tasks open more opportunities than others for enterprises not already supplying a particular organisation.

Straight Rebuys

Where a buyer, or purchasing department, routinely reorders an item it is referred to as a straight rebuy. The buyer will either reorder from the same supplier as last time or from an approved list of past suppliers. Entry for a new supplier is difficult and will possibly only be achieved if a potential supplier can either innovate in the form of new products or marketing systems (*e.g.* improved methods of transportation or materials handling), or can offer terms and conditions which are substantially better than the organisation is receiving from existing suppliers. The difference has to be substantial since any change of supplier involves risks and the buying organisation is unlikely to make a switch to achieve marginal gains.

Modified Rebuy

On occasion the buyer will want to modify the product specifications and/or the terms and conditions attached to a sale contract. Sometimes this opens up opportunities for enterprises which have not previously been suppliers to the buyer's organisation. New suppliers may be able to offer product specifications or terms that existing suppliers cannot match. Again, however, differences must be significant or buying organisations may not be sufficiently well motivated to accept the risks inherent in switching suppliers.

New Buy Task

Periodically buyers are faced with buying a product for the first time. The buyer's information search will almost certainly be more extensive. Since new buying tasks carry inherent risks, the buyer is likely to consult more widely with his/her colleagues and other advisers. This is where potential new suppliers, who are genuinely able to offer a better service, products and/or terms and conditions will find it easiest to gain access to buyers within the buying organisation.

According to Webster and Wind the decisions of industrial buyers are influenced by four groups of factors: environmental, organisational, interpersonal and individual. These are explained a little by Figure 4.5.

In summary, industrial marketers require a clear understanding of the needs of their prospective clients, an awareness of who is involved in the DMU, a knowledge of the criteria used in making buying decisions and an appreciation of the buying procedures involved.

Reseller Markets

Wholesalers, traders, sales agents, retailers and the like procure goods and services to resell or rent them to others, at a profit. These individuals and organisations comprise the reseller market. Some vertically integrated agricultural organisations are both producers and resellers. For instance, multinational fruit companies such as United Fruit, Geest, Del Monte own plantations as well as engaging in trading.

It is important that organisations wishing to market products and services into these markets understand that resellers act as purchasing agents on behalf of their customers and not as intermediaries for producers, manufacturers and other types

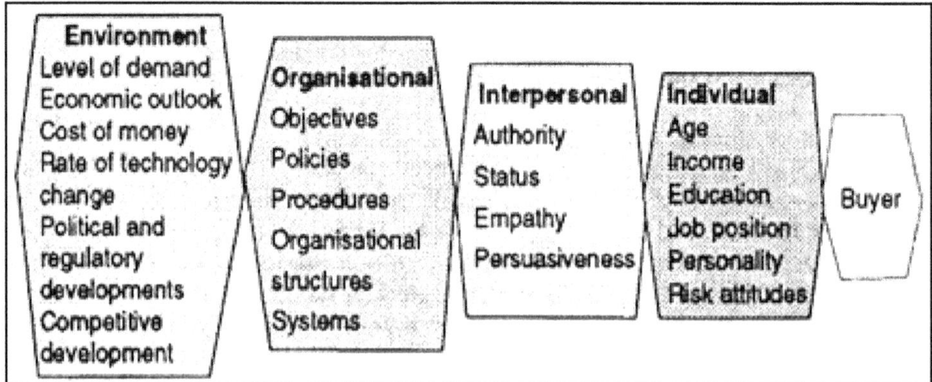

Figure 4.5: Major Influences on Organisational Buying Decisions.

of suppliers. The key to success in marketing to resellers is to seek to assist resellers in better meeting the needs of their customers.

Government Markets

Government markets have at least four levels. These are national, state, district and municipal. In most, if not all, countries the government is the largest single customer for goods and services. Government buys to fulfill mandated public objectives. Examples of the kinds of public objectives that involve the food and agriculture sectors are the maintenance of a strategic grain reserve, the procurement and distribution of staple foods to the poor and those suffering hardship due to drought, the management of nutrition intervention programmes for malnourished children, the provision of food for the country's armed forces and, sometimes, for schools, hospitals, prisons and other public institutions.

Government buying procedures usually take the form of either open bids or negotiated contracts. Open-bid buying involves a government agency in inviting tenders from pre-approved suppliers for carefully specified products, materials, works or services. In most cases, the government agency will also specify the terms and conditions attached to the contract. The normal practice is for the contract to be awarded to the lowest bidder able to meet the specifications. Prospective suppliers must decide both whether they can meet the specifications and if they are able to accept the terms and conditions laid down by the government agency.

Industrial Buyer Characteristics

Individuals who purchase products on behalf of an enterprise they either own or are employed by have two distinct sets of goals that they pursue: their own and those of the organisation. As an individual, the industrial buyer enjoys exercising authority, seeks job satisfaction, the approval and respect of both peers and superiors and other personal goals and avoids unnecessarily risky decisions. Industrial buyers are also motivated by the desire to achieve organisational goals such as cost control, improved efficiency of operations, reliable supplies of essential inputs, improved product performance and so on.

Figure 4.6: Motivations of Organisational Buyers.

Those marketing to organisational buyers can use their understanding of these twin sets of motivations to great effect.

Summary

The purpose of studying the social and psychological forces impinging upon customers is not to manipulate buyer behaviour but to be in a position to tailor the marketing mix so as to better meet the needs of prospective customers.

Buyer behaviour is shaped by social and psychological influences some of which are exogenous whilst others are endogenous. Exogenous influences, *i.e.* those arising from social interactions, are external to the individual, and include; culture, social status, reference groups, family membership and peer groups. Endogenous influences, *i.e.* those arising from the individual's own psyche and therefore internal, include needs and motives, perceptions, learning processes, attitudes and beliefs, personality and self-concept. It is only by understanding these sociological and psychological phenomena that an enterprise can really hope to develop a marketing mix that meets the needs of its target customers.

Chapter 5
Marketing Mix

Marketing Mix

Marketing mix is one of the major concepts in modern marketing. It is the combination of various elements which constitutes the company's marketing system. It is set of controllable marketing variables that the firm blends to produce the response it wants in the target market. Though there are many basic marketing variables, it is Mc Carthy, who popularized a four-factor classification called the four Ps: Product, Price, Place and Promotion. Each P consists of a list of particular marketing variables. Marketers often refer to the "Four *Ps*," or the marketing portfolio, as a way to describe resources available to market a product.

The first P – Product consists of

1. Product planning and development;
2. Product mix policies and strategies; and
3. Branding and packaging strategies.

The second P – Price consists of

1. Pricing policies and objectives; and
2. Methods of setting prices.

The third P – Place consists of

1. Different types of marketing channels;
2. Retailing and wholesaling institutions; and
3. Management of physical distribution systems.

The fourth P – Promotion consists of

1. Advertising;

2. Sales promotion; and
3. Personal selling.

When marketing their products firms need to create a successful mix of:

- the right product
- sold at the right price
- in the right place
- using the most suitable promotion.

To create the right marketing mix, businesses have to meet the following conditions:

- The product has to have the right features - for example, it must look good and work well.
- The price must be right. Consumer will need to buy in large numbers to produce a healthy profit.
- The goods must be in the right place at the right time. Making sure that the goods arrive when and where they are wanted is an important operation.
- The target group needs to be made aware of the existence and availability of the product through promotion. Successful promotion helps a firm to spread costs over a larger output.

For example, a company like Kellogg's is constantly developing new breakfast cereals - the product element is the new product itself, getting the price right involves examining customer perceptions and rival products as well as costs of manufacture, promotion involves engaging in a range of promotional activities *e.g.* competitions, product tasting etc, and place involves using the best possible channels of distribution such as leading supermarket chains.

A detailed discussion on each of the above four P's follows now:

Product

Product stands for various activities of the company such as planning and developing the right product and/or services, changing the existing products, adding new ones and taking other actions that affect the assortments of products. Decisions are also required in the are as such as quality, features, styles, brand name and packaging.

A product is something that must be capable of satisfying a need or want, it includes physical objects, services, personalities places, organization and ideas. Thus, a transport service, as it satisfiers human need is a product. Similarly, places like Kashmir and Kodaikanal, as they satisfy need to enjoy the cool climate are also products.

The second aspect of product is product planning and development. Product planning embraces all activities that determine a company's products. It include

1. Planning and developing a new product;
2. Modification of existing product lines; and
3. Elimination of unprofitable items.

Product development encompasses the technical activities of product research, engineering and decision. The product is the central point on which marketing energy must focus. Marketing therefore plays a key role in determining such aspects as:

- ✰ The appearance of the product - in line with the requirements of the market
- ✰ The function of the product - products must address the needs of customers as identified through market research.

The third aspect of product is product mix policies and strategies. Product mix refers to the composite of products offered for sale by a company. For example Godrej company offers cosmetics, steel furnitures, office equipments, locks *etc.* with many items in each category. The product mix is four dimensional. It has breadth, length, depth and consistency. Yet another integral part of product is packaging.

The product range and how it is used is a function of the marketing mix. The range may be broadened or a brand may be extended for tactical reasons, such as matching competition or catering for seasonal fluctuations. Alternatively, a product may be repositioned to make it more acceptable for a new group of consumers as part of a long-term plan.

Price

The second element of marketing mix is price. Price stands for the monetary value that customers pay to obtain the product. In pricing, the company must determine the right price for its products and then decide on strategies concerning retail and wholesale prices, discounts, allowances and credit terms. Of all the aspects of the marketing mix, price is the one, which creates sales revenue -all the others are costs. The price of an item is clearly an important determinant of the value of sales made. In theory, price is really determined by the discovery of what customers perceive is the value of the item on sale. Researching consumers' opinions about pricing is important as it indicates how they value what they are looking for as well as what they want to pay. An organization's pricing policy will vary according to time and circumstances.

Before fixing prices for the product, the company should be clear about its pricing objectives and strategies. The objectives may be set low initial price and raising it gradually or set high initial price and reducing it gradually or fixing a target rate of return or setting prices to meet the competition *etc.* But the actual price setting is based on three factors namely cost of production, level of demand and competition.

Different strategies may be taken with respect to price. Generically, there are two ways to make a profit—sell a lot and make a small margin on each unit or make a large margin on each unit and settle for lesser volumes. Firms in most markets are better off if the market is balanced—where some firms compete on price and others on other features (such as different taste preferences for different segments). The same idea applies at the retail level where some retailers compete on price while others compete on service while charging higher prices.

Regarding retail pricing, the company may adopt two policies. One policy is that he may allow the retailers to fix any price without interfering in his right. Another policy is that he may want to exercise control over the products. Discounts and allowances result in a deduction from the base price.

Place

The third element of marketing mix is place or physical distribution. Place stands for the various activities undertaken by the company to make the product accessible and available to target customers. Although figures vary widely from product to product, roughly a fifth of the cost of a product goes on getting it to the customer. 'Place' is concerned with various methods of transporting and storing goods, and then making them available for the customer. Getting the right product to the right place at the right time involves the distribution system. The choice of distribution method will depend on a variety of circumstances. It will be more convenient for some manufacturers to sell to wholesalers who then sell to retailers, while others will prefer to sell directly to retailers or customers.

There are four different level channels of distribution. The first is zero-level channel which means manufacture directly selling the goods to the consumers.

The second is one-level channel which means supplying the goods to the consumer through the retailer. The third is two-level channel which means supplying the goods to the consumer through wholesaler and retailer. The fourth is three-level channel which means supplying goods to the consumers through wholesaler-jobber-retailer and consumer.

There are large-scale institutions such as departmental stores, chain stores, mail order business, super-market *etc.* and small-scale retail institutions such as small retail shop, automatic vending, franchising *etc.* The company must chose to distribute their products through any of the above retailing institutions depending upon the nature of the products, area of the market, volume of scale and cost involved.

The actual operation of physical distribution system requires company's attention and decision-making in the areas of inventory, location of warehousing, materials handling, order processing and transportation.

Promotion

The fourth element of the marketing mix is promotion. Promotion stands for the various activities undertaken by the company to communicate the merits of its products and to persuade target customers to buy them. Advertising, sales promotion and personal selling are the major promotional activities. A perfect coordination among these three activities can secure maximum effectiveness of promotional strategy. The cost associated with promotion or advertising goods and services often represents a sizeable proportion of the overall cost of producing an item. However, successful promotion increases sales so that advertising and other costs are spread over a larger output. Though increased promotional activity is often a sign of a response to a problem such as competitive activity, it enables an organization to develop and build up a succession of messages and can be extremely cost-effective.

Chapter 6

Marketing Functions

Any single activity performed in carrying a product from the point of its production to the ultimate consumer may be termed as a marketing function. A marketing function may have anyone or combination of three dimensions, *viz.*, time, space and form.

The marketing functions involved in the movement of goods from the producer to its ultimate consumer vary from commodity to commodity, market to market, the level of economic development of the country or region, and the final form of the consumption.

The marketing functions may be classified in various ways. Thomsen has classified the marketing functions into three broad groups.

1. Primary Functions

- ☆ Assembling or procurement
- ☆ Processing
- ☆ Dispersion or Distribution

2. Secondary Functions

- ☆ Packing or Packaging
- ☆ Transportation
- ☆ Grading, Standardization and
- ☆ Quality Control
- ☆ Storage and Warehousing
- ☆ Price Determination or Discovery
- ☆ Risk Taking
- ☆ Financing

- ☆ Buying and Selling
- ☆ Demand Creation
- ☆ Dissemination of Market Information

3. Tertiary Functions

- ☆ Banking
- ☆ Insurance
- ☆ Communications
- ☆ Supply of Energy – Electricity

Kohls and Uhl have classified marketing functions as follows:

1. Physical Functions

- ☆ Storage and Warehousing
- ☆ Grading
- ☆ Processing
- ☆ Transportation

2. Exchange Functions

- ☆ Buying
- ☆ Selling

3. Facilitative Functions

- ☆ Standardization of grades
- ☆ Financing
- ☆ Risk Taking
- ☆ Dissemination of Market Information

Packaging

Packaging is the first function performed in the marketing of agricultural commodities. It is required for nearly all farm products at every stage of the marketing process. The type of the container used in the packing of commodities varies with the type of the commodity as well as with the stage of marketing. For example, gunny bags are used for cereals, pulses and oilseeds when they are taken from the farm to the market. For packing milk or milk products, plastic, polythene, tin or glass containers are used.

Wooden boxes, straw baskets CFB boxes and carates are used for packing fruits and vegetables.

Meaning of Packing and Packaging

Packing means, the wrapping and crating of goods before they are transported. Goods have to be packed either to preserve them or for delivery to buyers. Packaging is a part of packing, which means placing the goods in small packages like bags, boxes, bottles or parcels for sale to the ultimate consumers. In other words, it means putting goods on the market in the size and pack which are convenient for the buyers.

Advantages of Packing and Packaging

Packaging is a very useful function in the marketing process of agricultural commodities. Most of the commodities are packed with a view to preserving and protecting their quality and quantity during the period of transit and storage. For some commodities, packing acts as a powerful selling tool. The chief advantages of packing and packaging are:

- It protects the goods against breakage, spoilage, leakage or pilferage during their movement from the production to the consumption point.
- The packaging of some commodities involves compression, which reduces the bulk like cotton, jute and wool.
- It facilitates the handling of the commodity, specially such fruits as apples, mangoes, *etc.*, during storage and transportation.
- It helps in quality-identification, product differentiation, branding and advertisement of the product, *e.g.*, Safal peas and Amul butter.
- Packaging helps in reducing the marketing costs by reducing the handling and retailing costs.
- It helps in checking adulteration.
- Packaging ensures cleanliness of the product.
- Packaging with labeling facilitates the conveying of instructions to the buyers as to how to use or preserve the commodity. The label shows the composition of the product.
- Packaging prolongs the storage quality of the products by providing protection from the ill effects of weather, specially for fruits, vegetables and other perishable goods.

Transportation

Transportation or the movement of products between places is one of the most important marketing functions at every stage, *i.e.*, right from the threshing floor to the point of consumption. Most of the goods are not consumed where they are produced. All agricultural commodities have to be brought from the farm to the local market and from there to primary wholesale markets, secondary wholesale markets, retail markets and ultimately to the consumers. The inputs from the factories must be taken to the warehouses and from the warehouses to the wholesalers, retailers and finally to the consumers (farmers). Transportation adds the place utility to goods.

Transport is an indispensable marketing function. Its importance has increased with urbanization. Trade and transport go side by side; the one reinforces and strengthens the other.

Advantages of Transport Function

The main advantages of the transport function are:

- **Widening of the Market:** Transport helps in the development or widening of markets by bridging the gap between the producers and consumers located in different areas. The exchange of goods between different districts, regions or countries would be impossible in the absence of this function.
- **Narrowing Price Difference Over Space:** The transportation of goods from surplus areas to the places of scarcity helps in checking price rises in the scarcity areas and price falls in surplus areas thus reduces the spatial differences in prices.
- **Facilitation of Specialized Farming:** Different areas of the country are suitable for different crops, depending on their soil and agroclimatic conditions. Farmers can go in for specialization in the commodity most suitable to their area, and consumers exchange the goods required by them from other areas at a cheaper price than their own production cost.
- **Transformation of the Economy:** Transportation helps in the transformation of the economy from the subsistence stage to the developed commercial stage. Industrial growth is stimulated by being fed with the raw material produced in rural areas. Manufactured goods from industries to village or rural areas, too, can be moved.
- **Mobility of the Factors of Production:** Transport helps in increasing the mobility of capital and labour from one area to another. Entrepreneurs get opportunities for the investment of their capital in newly opened areas of the country, where the prospects of profit are very bright. Moreover, transportation helps in the migration of people in search of better remunerative jobs.

Transportation Cost

The transportation cost accounts for about 50 percent of the total cost of marketing. The efficiency of transportation depends on the speed and the care with which goods move from one place to another, the extent of the facilities provided, and the degree of care with which goods are handled en route and at terminal stations. However, there is a need for reducing the cost of transportation.

Factors Affecting the Cost of Transportation

The transportation cost of a commodity depends on the following factors:

- **Distance:** With an increase in the distance over which a commodity is transported, the total transportation cost increases; but the transportation cost per unit quantity of the produce decreases after a certain distance.
- **Quantity of the Product:** The transportation cost per unit quantity of a commodity decreases with the increase in the volume. It will be less if a full truckload is available than it would be if only a few quintals are transported.

- ☆ **Mode of Transportation:** The cost of transportation varies with the mode of transportation, *e.g.*, bullock cart, tractor, truck, railway *etc.*
- ☆ **Condition of Road:** The cost of transportation is less where metalled or tar roads have been constructed than in places where graveled roads exist or where there are no roads at all.
- ☆ **Nature of Products:** The cost of transportation per unit is higher for the products having the following characters:
  - ❒ Perishability (*e.g.*, Vegetables);
  - ❒ Bulkinesss (*e.g.*, straw);
  - ❒ Fragility;
  - ❒ Inflammability (*e.g.*, Petrol);
  - ❒ Requirement of a special type of facility (for example, for livestock and milk).
- ☆ **Availability of Return Journey consignment:** If goods are also available for transportation when a truck is to return to its starting place, the per unit cost of transportation is less.
- ☆ **Risk Associated:** The transportation cost is less if the produce is transported at the owners/senders risk than when the risk is on the agency transporting the produce.

Problems in Transportation of Agricultural Commodities

The problems in the transportation of agricultural commodities are very serious because of the special factors associated with them. The following are some of the important problems arising out of the transportation of agricultural commodities:

- ☆ There are more losses/damages in transportation because of the use of poor packaging material, overloading of the produce and poor handling, specially of fruits and vegetables, at the time of loading and unloading;
- ☆ The transportation cost of the farm produce is higher than that for other goods. This is so because of its bulky character and the prevailing practice of fixing charges on the basis of weight or volume rather than on the basis of its value;
- ☆ There is lack of co-ordination between different transportation agencies, *e.g.*, the railways and truck companies. Some of the places are not connected by railway.
- ☆ The produce is often transported for a part of the distance by rail and a part by trucks or other means of transportation.

Suggestions for Improvement

There must be full utilization of the capacity of the transportation facility in terms of the load. This would reduce the per quintal cost of transportation.

The transportation cost per quintal can be reduced by fixing the rate of transportation for different means. At present, each agency charges what it likes and not on the basis of any rational computation of the cost factor.

There should be a reduction in spoilage, damage, breakage and pilferage during the period of movement as a result of better handling, packing and the use of the proper types of wagons.

There should be a reduction in the barriers to inter-state movement of the produce. If this happens, the time taken in transportation and the quantity of the fuel consumed would be reduced.

A reduction in the bulk of the produce by processing it can help in minimizing the transport cost. For example, milk may be processed into condensed milk, butter or ghee and fruits into juices.

The speed and capacity of the vehicles used in transportation should be increased.

Nearly 50 percent of the villages in the country are still not connected by roads. This apart, there are sharp differences among the states. The rail transport, though capable of transporting agricultural commodities to longer distances in larger quantities with greater speed but it also suffers from multi-gauge system, shortages of wagon capacity and congestion on trunk routes. Therefore, in the overall scheme of public investment, development of this infrastructure should receive more allocation.

Grading and Standardization

Grading and standardization is a marketing function which facilitates the movement of produce. Without standardization the rule of caveat emptor (let the buyer beware) prevails; and there is confusion and unfairness as well. Standardization is a term used in a broader sense. Grade standards for commodities are laid down first and then the commodities are sorted out according to the accepted standards.

Products are graded according to quality specifications. But if these quality specifications vary from seller to seller, there would be a lot of confusion about its grade.

The top grade of one seller may be inferior to the second grade of another. This is why buyers lose confidence in grading. To avoid this eventuality, it is necessary to have fixed grade standards which are universally accepted and followed by all in the trade.

Meaning

Standardization means the determination of the standards to be established for different commodities. Pyle has defined standardization as the determination of the basic limits on grades or the establishment of model processes and methods of producing, handling and selling goods and services.

Standards are established on the basis of certain characteristics-such as weight, size, colour, appearance, texture, moisture content, staple length, amount of foreign matter, ripeness, sweetness, taste, chemical content, *etc.* These characteristics, on the basis of which products are standardized, are termed grade standards. Thus, standardization means making the quality specifications of the grades uniform among buyers and sellers over space and over time.

Grading means the sorting of the unlike lots of the produce into different lots according to the quality specifications laid down. Each lot has substantially the same characteristics in so far as quality is concerned. It is a method of dividing products into certain groups or lots in accordance with predetermined standards. Grading follows standardization. It is a sub-function of standardization.

Advantages of Grading

Grading offers the following advantages to different groups of persons:

- Grading before sale enables farmers to get a higher price for their produce.
- Grading also serves as an incentive to producers to market the produce of better quality.
- Grading facilitates marketing, for the size, color, qualities and other grade designations of the product are well known to both the parties, and there is no need on the part of the seller to give any assurance about the quality of the product.
- Grading widens the market for the product, for buying can take place between the parties located at distant places on the telephone without any inspection of the quality of the product.
- Grading reduces the cost of marketing by minimizing the expenses on the physical inspection of the produce, minimizing storage loses, reducing its bulk, minimizing advertisement expenses and eliminating the cost of handling and weighing at every stage.

Grading makes it possible for the farmer:

- To get easy finance when commodities are stored;
- To get the claims settled by the railways and insurance companies;
- To get storage place for the produce;
- To get market information;
- To pool the produce of different farmers;
- To improve the "keeping" quality of the stored products by removing the inferior goods from the good lot; and
- To facilitate futures trading in a commodity.

Grading helps consumers to get standard quality products at fair prices. It is easier for them to compare the prices of different qualities of a product in the market. It minimizes their purchasing risk, for they will not get a lower quality product at the given price.

Thus, the grading of product is beneficial to all the sections of society; *i.e.*, the producers, traders and consumers of the product.

In India to improve the quality of agricultural products in India, grading and marking were introduced under an Act – The Agricultural Produce (Grading and Marking) Act, 1937.The act authorizes the Central Government to frame rules relating to the fixing of grade standards and the procedure to be adopted to grade the agricultural commodities included in the schedule. Initially, only 19 commodities were included for grading purposes; but now there are 153 commodities in the schedule for which grade standards are available. The commodities included in the schedule are food grains, fruits and vegetables, dairy products, tobacco, coffee, oilseeds, edible oils, oilcakes, fruit products, cotton, sanhemp, edible nuts, jaggery, lac, spices and condiments, essential oils, honey, besan, suji and maida.

The Agricultural Marketing Advisor to the Government of India (AMA) is the authority empowered to implement the provisions of the Act, and suggest suitable modifications. The Central Agricultural Marketing Department (Directorate of Marketing and Inspection) maintains some staff for the inspection of the grading premises and the collection of the samples of graded products from different points in the marketing process. The collected samples are examined and analyzed either at the central Agmark laboratory or at other laboratories set up in different parts of the country to test whether the graded products conform to the standards of quality laid down in the Act. If the sample is below standard, the necessary legal action against the party is taken, and the graded product is removed from the market. The license of the party, too, is cancelled.

Storage

Meaning and Need

Storage is an important marketing function, which involves holding and preserving goods from the time they are produced until they are needed for consumption. Storage is an exercise of human foresight by means of which commodities are protected from deterioration, and surplus supplies in times of plenty are carried over to the season of scarcity. The storage function, therefore, adds the time utility to products.

Agriculture is characterized by relatively large and irregular seasonal and year – to – year fluctuations in production. The consumption of most farm products, on the other hand, is relatively stable. These conflicting behaviors of demand and supply make it necessary that large quantities of farm produce should be held for a considerable period of time.

The storage function is as old as man himself, and is performed at all levels in the trade. Producers hold a part of their output on the farm. Traders store it to take price advantage. Processing plants hold a reserve stock of their raw materials to run their plants on a continuous basis. Retailers store various commodities to satisfy the consumers day – to –day needs. Consumers, too, store food grains, depending on their financial status.

The storage of agricultural products is necessary for the following reasons:

- ✰ Agricultural products are seasonally produced, but are required for consumption throughout the year. The storage of goods, therefore, from the time of production to the time of consumption, ensures a continuous flow of goods in the market;

- ☆ Storage protects the quality of perishable and semi – perishable products from deterioration;
- ☆ Some of the goods, have a seasonal demand. To cope with this demand, production on a continuous basis and storage become necessary. It helps in the stabilization of prices by adjusting demand and supply;
- ☆ Storage is necessary for some period for the performance of other marketing functions. For example, the produce has to be stored till arrangements for its transportation are made, or during the process of buying and selling, or the weighment of the produce after sale, and during its processing by the processor;
- ☆ The storage of some farm commodities is necessary either for their ripening (*e.g.*, banana, mango, *etc.*) or for improvement in their quality (*e.g.*, rice, pickles, tobacco, *etc.*)
- ☆ Storage provides employment and income through price advantages. For example, middlemen store food grains by purchasing them at low prices in the peak season and sell them in the other seasons when prices are higher.

Risks in Storage

The storage of agricultural commodities involves three major types of risks. These are:

- ☆ **Quantity Loss:** The risks of loss in quantity may arise during storage as a result of the presence of rodents, insects and pests, theft, fire, *etc.* Dehydration too, brings about an unavoidable loss in weight. It has been estimated that about 10 million tones of foodgrains are lost every year because of poor and faulty storage.
- ☆ **Quality Deterioration:** The second important risk involved in the storage of farm products is the deterioration in quality, which reduces the value of the stored products. These losses may arise as a result of attack by insects and pests, the presence of excessive moisture and temperature, or as a result of chemical reaction during the period of storage. Dehydration of fruits, vegetables and meat during storage may lower their sale value. The loss in the quality of farm products varies with their quality at the time of storage, the method of storage and the period of storage.
- ☆ **Price Risk:** This, too, is an important risk involved in the storage of farm products. Prices do not always rise enough during the storage period to cover the storage costs. At times they fall steeply, involving the owner in a substantial loss. Farmers and traders generally store their products in anticipation of price rise and they suffer when prices fall.

Cold Storage

"Cold storage warehouse" shall mean any place artificially cooled to or below a temperature above zero of 45 degrees Fahrenheit in which articles of food are placed and held for thirty days or more.

- ✰ First cold storage warehouse established in New York in 1865 for fish
- ✰ In India cold storage order was passed in 1964. In India the first cold storage was established in 1892 at Calcutta.

Agencies Involved

In India, the Central Warehousing Corporations (CWC), State Warehousing Corporations (SWC) and Food Corporation of India (FCI) are involved in storing the major agri-commodities. The private parties like ITC, Cash and Carry are coming up in this sector. The government is also encouraging the private parties to participate in this process.

Processing

Processing is an important marketing function in the present-day marketing of agricultural commodities. A little more than 100 years ago, it was a relatively unimportant function in marketing. A large proportion of farm products were sold in an unprocessed form, and a great deal of the processing was done by the consumers themselves. At present, consumers are dependent upon processing for most of their requirements. Many technological changes have occurred in the recent past, such as the introduction of refrigeration, modern methods of milling and baking food grains, new processing methods for dairy products, and modern methods of packing and preservation.

These technological changes have had a significant impact on the standard of living of the consumers, on the economic and social organizations of society, and on the growth of trade in the country.

Meaning

The processing activity involves a change in the form of the commodity. This function includes all of those essentially manufacturing activities which change the basic form ofthe product. Processing converts the raw material and brings the products nearer to human consumption. It is concerned with the addition of value to the product by changing its form.

Advantages

The processing of agricultural products is essential because very few farm products –milk, eggs, fruits and vegetables – are consumed directly in the form in which they are obtained by the producer. All other products have to be processed into a consumable form. Processing is important, both for the producer – sellers and for consumers. It increases the total revenue of the producer by regulating the supply against the prevailing demand. It makes it possible for the consumer to have articles in the for mliked by him. The specific advantages of the processing function are:

- ☆ It changes raw food and other farm products into edible, usable and palatable forms. The value added by processing to the total value produced at the farm level varies from product to product. It is nearly 7 percent for rice and wheat, about 79 percent for cotton and 86 percent for tea. It is generally higher for commercial crops than for food crops. Examples of the products in this group are: the processing of sugarcane to make sugar, gur, khandsari; oilseeds processing to make oil; grinding of food grains to make flour; processing of paddy into rice; and conversion of raw mango into pickles.
- ☆ The processing function makes it possible for us to store perishable and semi-perishable agricultural commodities which otherwise would be wasted and facilitates the use of the surplus produce of one season in another season or year.
- ☆ Examples of the processing of the products in this group are: drying, canning and pickling of fruits and vegetables, frozen goods, conversion of milk into butter, ghee and cheese and curing of meat with salting/smoking.
- ☆ The processing activity generates employment. The baking industry, the canning industry, the brewing and distilling industry, the confectionary industry, the sugar industry, oil mills and rice mills provide employment to a large section of society.
- ☆ Processing satisfies the needs of consumers at a lower cost. If it is done at the door of the consumer, it is more costly than if it is done by a firm on large scale.
- ☆ Processing saves the time of the consumers and relieves them of the difficulties and botherations experienced in processing.
- ☆ Processing serves as an adjunct to other marketing functions, such as transportation, storage and merchandising.
- ☆ Processing widens the market. Processed products can be taken to distant and overseas markets at a lower cost.

Buying and Selling

Meaning

Buying and selling is the most important activity in the marketing process. At every stage, buyers and sellers come together, goods are transferred from seller to buyer, and the possession utility is added to the commodities.

The number of times the selling-and-buying activity is performed depends on the length of the marketing channel. In the shortest channel where no middleman is involved, this activity takes place only once, *i.e.*, the producer or farmer sells and the consumer purchases. But, usually, in the case of farm commodities, selling/buying activities are undertaken each time when the produce moves from the farmer to the primary wholesaler, from the wholesaler to the retailer, and from the retailer to the consumer.

The buying activity involves the purchase of the right goods at the right place, at the right time, in the right quantities and at the right price. It involves the problems of what to buy, when to buy, from where to buy, how to buy and how to settle the prices and the terms of purchase.

The selling activity involves personal or impersonal assistance to or persuasion of, a prospective buyer to buy a commodity. The objective of selling is to dispose of the goods at a satisfactory price. The prices of products, particularly of agricultural commodities vary from place to place, from time to time, and with the quantity to be sold. Selling, therefore, involves the problems of when to sell, where to sell, through whom to sell, and whether to sell in one lot or in parts.

Methods

The following methods of buying and selling of farm products are prevalent in Indian markets:

Under Cover of a Cloth (Hatha System)

By this method, the prices of the produce are settled by the buyer and the commission agents of the seller by pressing/twisting the fingers of each other under cover of a piece of cloth. Code symbols are associated with the twisting of the fingers, and traders are familiar with these. This system provides opportunities for cheating the seller, for the seller is not aware of the price that has been offered by other buyers; the commission agent may not communicate the various prices to the seller, and may strike a deal in favour of one who offers a somewhat lower price. This method has been banned by the government because of the possibility of cheating, though it continues to be used in some markets.

Private Negotiations

By this method, prices are fixed by mutual agreement. This method is common in unregulated markets or village markets. The sellers takes the sample to the buyer and asks him to quote the price. If it is acceptable to the buyer, a contract is executed. This however, is a slow and time-consuming process and is not suitable when either large quantities have to be sold or a large number of buyers exist in the market. The advantage of this method is that the seller gets a good price, for buyers are not aware of the price offered by other buyers. Each buyer, therefore, tries to bid the highest to get the produce.

Quotations on Samples taken by Commission Agent

By this method the commission agent takes the sample of the produce to the shops of the buyer instead of the buyer going to the shop of the commission agent. The price is offered, based on the sample, by the prospective buyers. The commission agent makes a number of rounds of prospective buyers until none is ready to bid a price higher than the one offered by a particular buyer. The produce is given to the one whose bid has been the highest.

Dara Sale Method

By this method, the produce in different lots is mixed and then sold as one lot. The advantages of this method is that, within a short time, a large number of lots are sold off. The disadvantage is that the produce of a good quality and one of a poor quality fetch the same price. There is, therefore, a loss of incentive to the farmer to cultivate good quality products. This method is common for such crops as zeera in many markets of the country.

Moghum Sale Method

By this method, the prospective buyers gather at the shop of the commission agent around the heap of the produce, examine it and offer bids loudly. The produce is given to the highest bidder after taking the consent of the seller farmer. This method is preferred to any other method because it ensures fair dealing to all parties, and because the farmers with a superior quality of produce receive a higher price. In most regulated markets, the sale of the produce is permissible only by the open auction method.

Open Auction Method

By this method, the prospective buyers gather at the shop of the commissionagent around the heap of the produce, examine it and offer bids loudly. Theproduce is given to the highest bidder after taking the consent of the sellerfarmer. This method is preferred to any other method because it ensures fairdealing to all parties, and because the farmers with a superior quality ofproduce receive a higher price. In most regulated markets, the sale of theproduce is permissible only by the open auction method.

The following are the merits of the open auction method:

- A sale by this method inspires confidence among the buyers and sellers.
- The seller is able to follow the bidding easily.
- The auction serves as a meeting place for the supply of, and demand for goods.
- It disposes of the market supply promptly.
- A wide variety of goods are available to buyers for selection.
- The auction method reduces the number of salesmen needed in the process.
- The buyers of small lots are not put to a disadvantage against the buyers of large lots.
- All the sections interested in the sale and purchase are well informed about the prevailing prices and can take judicious decisions about the sale and purchase of agricultural commodities.
- The payment of the price of the goods is made immediately after the sale if an auction has been completed.

The disadvantages of the open auction method are:

- ☆ The auction method requires more time on the part of both the buyer and the seller, for they have to wait for the day and time of the auction. An open auction is a very time-consuming process because of the variation in the quality of the various lots.
- ☆ In big market centers, specially in the peak marketing season, the time allotted for auction is short. Both the buyers and the sellers are in a hurry. As a result, sellers may receive a low price.
- ☆ In an open auction, buyers sometimes join hands. Active participation init is then reduced.
- ☆ The auction leads to a "buyers market", for buyers have full information about the supply of, and demand for the product.
- ☆ Some of the problems arising out of the open auction method may be overcome if the grading of agricultural produce is adopted by the cultivators. This will reduce the time involved in inspection and bidding for each lot separately, and will result in increasing the overall efficiency of the marketing system.

The methods employed for the sale of agricultural commodities in Indian markets differ from market to market and also from commodity to commodity. However, in regulated markets, either the open auction or the close tender system is prevalent.

Market Information

Market information may be broadly defined as a communication or reception of knowledge or intelligence. It includes all the facts, estimates, opinions and other information which affect the marketing of goods and services. Market information is an important marketing function which ensures the smooth and efficient operation of the marketing system. Accurate, adequate and timely availability of market information facilitates decision about when and where to market products. Market information creates a competitive market process and checks the growth of monopoly or profiteering by individuals. Everyone engaged in production, and in the buying and selling of products is continually in need of market information. This is more true where agricultural products are concerned, for their prices fluctuate more widely than those of the products of other sectors. Market information is essential for the government, for a smooth conduct of the marketing business, and for the protection of all the groups of persons associated with this. It is essential at all the stages of marketing, from the sale of the produce at the farm until the goods reach the last consumer.

Market information helps in improving the decision-making power of the farmer. A farmer is required to decide when, where and through whom he should sell his produce and buy his inputs. Price information helps him to take these decisions.

Market middlemen need market information to plan the purchase, storage and sale of goods. The failure of a business may partly be attributed to either the non-availability of market information or its inadequate availability and interpretation. Market information is essential for the government in framing its agricultural policy relating to the regulation of markets, buffer stocking, import export, and administered prices.

Financing

There is a long interval between the time of production and consumption. Between these two points, the ownership of commodities shifts many times – a fact which necessitates financial arrangements. Middlemen need finance not only for the purchase of stocks, but for the performance of various marketing functions, such as processing, storage, packaging, transport and grading. The financing function of marketing involves the use of capital to meet the financial requirements of the agencies engaged in various marketing activities.

No business is possible nowadays without the financial support of other agencies because the owned funds available with the producers and market middlemen (such as wholesalers, retailers and processors) are not sufficient. The financial requirements increase with the increase in the price of the produce and the cost of performing various marketing services.

Factors affecting Capital Requirements of an Agricultural Marketing Firm

The capital requirement of a marketing agency for its marketing business varies with the following factors:

- ☆ Nature and Volume of Business: Financial requirements for trading in high value crops like cumin, chillies, cotton and oilseeds are higher than for trading in food grains. For the wholesale business too, financial requirements are higher than for retail business.
- ☆ Necessity of Carrying Large Stocks: It is essential to carry over large stocks throughout the year, of goods which are seasonally produced and marketed on a wholesale basis.
- ☆ Continuity of Business during Various Seasons: If business is continuous throughout the year, the financial requirements will be greater than if business is conducted only during a particular season.
- ☆ Time Required between Production and Sale: Some goods are sold immediately after production – perishables, for example fruits and vegetable while others are disposed for after a certain time – rice and cheese, for example. Financial requirements in the marketing of the latter goods are, therefore higher.
- ☆ Terms of Payment for Purchase and Sale: The terms of transactions – whether payment will be in cash, on credit or by installments – affect the financial requirements of the marketing middlemen.

- ☆ Fluctuations in Prices: Financial requirements are higher for goods which suffer frequent price fluctuations than for goods that are subject to less frequent price fluctuations.
- ☆ Risk-taking Capacity: The financial needs of the market middlemen vary with their risk-taking capacity. A middleman with a low risk-taking capacity often resorts to hedging, and needs less finance than a middleman who takes risks.
- ☆ General Conditions in the Economy: During the period of price fall or recession, the financial requirements increase. The marketing agency has to hold stocks for a longer period in anticipation of a price rise. Moreover, the recovery of old bills tends to be slow. Whenever, therefore, a new product is introduced, the dealer needs more finance temporarily till the demand for it picks up in the economy.

The marketing finance required by the marketing middlemen is of two types – fixed capital for land, buildings (shops and godowns), equipment and machinery(weighbridge, grading equipment, *etc.*), and working capital which is required to meet the marketing costs, purchase value, and salaries of the employees. The proportion of working capital is higher than that of fixed capital. It is also necessary to make arrangements for financing the farmers during the period between the production and sale of their produce. This is necessary to improve their holding capacity and to avoid the post-harvest sale of the produce when prices are low in the market.

Because of their acute financial needs, many farmers market their standing crops – of fruits, for example – or borrow money in advance from local traders/ commission agents against their crops, and bind themselves to sell the crop through the trader/commission agent. This checks their freedom to sell the produce in the open market.

To improve the financial position of the farmers and to strengthen their holding capacity, the following steps have been taken by the government.

- ☆ Commercial banks have started financing the agricultural sector in a big way and meeting the increasing needs of the farmers for production purposes.
- ☆ The co-operatives, too, have developed and entered the field of agricultural financing. Under this scheme, co-operative credit societies can realize their credit, together with the interest due on it, by the sale proceeds of the produce directly by intimation to Co-operative Marketing Societies. These may make the payments for the produce to the farmer after deducting their dues. A rapid progress has been made in this area.
- ☆ With the development of warehousing facilities in the country, farmers can now meet 70 to 80 percent of their credit needs by placing the produce in the warehouses. Banks extend the financing facility to farmers against the mortgage of the warehouse receipt. This scheme has lessened the financial problems of the farmers and of market middlemen. As a result,

the tendency to sell the produce immediately after the harvest should have been checked. However, it has met with only limited success. So long as the interest rate continues to be more than the intra-year rise in prices, storage cannot be a profitable proposition.

Chapter 7

Product Decisions

Product, the first of the four Ps of marketing mix has a unique position as it constitutes the most substantive element in any marketing offer. The other elements – price, place and promotion –are normally employed to make the product offering unique and distinct. Product is, thus, the number one weapon in the marketer's arsenal.

Product is complex concept which has to be carefully defined. In common parlance, any tangible items such as textiles, books, television and many others are called as products. But an individual's decision to buy an item is based on not only on its tangible attributes but also on a variety of associated non-tangible and psychological attributes such as services, brand, package, warranty, image *etc.* Therefore, to crystallize the understanding of the term 'product', it would be appropriate to take recourse to different definitions of 'product' given by marketing practioners. According to Alderson, "Product is a bundle of utilities consisting of various product features and accompanying service". The bundle of utilities is composed of those physical and psychological attributes that the buyer receives when the buys he product and which the marketer provides a particular combination of product features and associated services.

Philip Kotler defines product 'as anything that can be offered by a marketer for attention, acquisition, use consumption that might satisfy a want or need. It includes physical object, services persons, places organizations and ideas.

The perusal of above definitions reveal that a product is not only an tangible entity, but also the intangible services such as prestige, image *etc.* form an integral part of the product.

Precisely, the answers to the following question is the product policy of a firm:

- ☆ What products should the company make?
- ☆ Where exactly are these products to be offered?

- To which market or market segment?
- What should be the relationship among the various members of a product line?
- What should be the width of the product mix?
- How many different product lines can the company accommodate?
- How should the products be positioned in the market?
- What should be the brand policy?
- Should there be individual brands, family brands and/or multiple brands?

A product policy serves the following three main functions:

1. A product policy guides and directs the activities of whole organization toward a single goal. Only rarely, product decisions are made solely by top executives. More often such decisions require the specialized knowledge of experts in many fields –research, development, engineering, manufacturing, marketing, law, finance and even personnel.
2. A product policy helps to provide the information required for decisions on the product line.
3. A product policy gives executives a supplementary check on the usual estimates of profit and loss.

A sound product policy is thus an important tool for coordination and directions. It applies not only to those major decisions which are ultimate responsibility of general managers also to the many lower level employees who also take day to day decisions.

Product Classification

Marketers have developed several product classification schemes based on product characteristics as an aid to developing appropriate marketing strategies.

Product can be classified into three groups according to their durability:

1. **Durable Goods**: Durable goods are tangible goods that normally survive many uses. Examples include refrigerators, televisions, automobiles, *etc.*
2. **Non-Durable Goods**: These are tangible goods that normally are consumed for short period. Example include food products, toiletries, tractors, *etc.*
3. **Services:** Services are activities, benefits or satisfactions that are offered for sale. Examples include banking, transport, insurance service *etc.*

Another method of classifying products is on the basis of consumer shopping habits because they have implications for marketing strategy. Based on this, goods may be classified into three:

Convenience Goods

Goods that the customer usually purchases frequently, immediately and with the minimum effort. The price per unit is low, Example: soaps, match box *etc.*

Shopping Goods

These goods are purchased infrequently. The price per unit is comparatively higher. The customer, in the process of selection and purchase of these goods compares the suitability, quality, price and style. Example includes furniture, clothing, footwear *etc.*

Specialty Goods

Goods with unique characteristics and/or brand identification for which a significant group of buyers are willing to make a special purchasing effort. The goods are expensive and purchased rarely. Examples include personal computers, cars, jewellery *etc.*

Industrial Products

One of the ways of classification of industrial products involves two broad categories *viz.* (1) products that are used in the production of other goods and become a physical part of another product, and (2)products necessary to conduct business that do not become part of another product. The products that become part of another product are raw materials, semi-manufactured goods, components and subcontracted production services. The products that are needed to conduct the business include: Capital goods, operating supplies, contracted industrial services, contracted professional services and utilities.

Raw material include crude oil, coal, iron ore, other mined minerals, lumber, forestry product, agricultural products, livestock, poultry and dairy products and the products of fisheries.

Semi-manufacturing goods are products, that when purchased, have already undergone some processing but are incomplete in themselves. Examples are cotton fiber, castings, plate glass and plastics.

Components are completed products meant to become part of another larger, more complicated product. Examples include automobile batteries, headlights, tyres *etc.*

Subcontracted production Examples are, subcontracting for installation of electrical, heating, air-conditioning and plumbing facilities to others.

Capital goods are manufacturing plants and installations, tools, machines, trucks *etc.*

Operating supplies are industrial products used to keep a business operating normally. These include lubricating oils, paperclips, cash registers *etc.* The operating supplies usually have a relatively low unit value, and are consumed quickly.

Contracted industrial services include such items as machine servicing and repair, cleaning, remodeling, waste disposal and the operation of the employees' canteens. Contracted professional services include printing executive recruitment, advertisement, advertising, legal advice, professional accounting, data processing and engineering studies.

The industrial products in the category of utilities consist of energy, telephone, and water.

In order to further understand, it will be appropriate to know the meaning of some other terms also which often recur in any discussion about product. Some of these terms are discussed below:

Need Family: The core need that actualizes the product family. Example: Safety.

Product Family: All the product classes that can satisfy a core need with more or less effectiveness, ex. Beverages.

Product Line: A group of products within a product class that are closely related, because they function in a similar manner or sold to the same customer groups or are marketed through the same types of outlets or fall within given price ranges. Example: Cosmetics, Agriinputs.

Product Item: A distinct unit within a brand or product line that is distinguishable by size, price, appearance or some other attribute. Example: Talcum powder.

Product Mix Decisions

Product Mix

A product mix (also called product assortment) is the set of all product lines and items that a particular seller offers to sale.

A company's product mix can be described as having a certain width, length, depth, and consistency.

The width of the product mix refers to how many product lines the company carries.

The length of product mix refers to the total number of items in its product mix.

The depth of product mix refers to how many product variants are offered of each product item in the line.

The consistency of the product mix refers to how closely related the various product lines are in end use, product requirements, distribution channels or some other way.

These four dimensions of the product mix provide the bases for defining the company's product strategy. The company can grow its business in four ways. The company can add new product lines, thus widening its product mix to capitalize the company's reputation or the company can lengthen its existing product lines to become a more full line company or the company can add more product variants to each product and thus deepen its product mix. Finally the company can pursue more product-line consistency or less, depending upon whether it wants to acquire a strong reputation in a single field or participate in several fields.

Product Line Decisions

Product Line

A product line is a group of products that are closely related, because they function in a similar manner, are sold to the same customer groups, are marketed through the same types of outlets, or fall within given price ranges.

Product line managers have two important information needs. First they must know the sales and profits of each item in the line. Second, they must know how the product line compares to competitor's product lines in the same markets (Product Positioning).

One of the major issues facing product-line managers is the optimal length of the product line. The manager can increase the profits either by adding the product items if the line is too short or by dropping the items if the line is too long.

The issue of product-line length is influence by company objectives. Companies that want to be positioned as full-lines companies and/or are seeking high market share and market growth will carry longer lines.

They are less concerned when some items fail to contribute to profit. Companies that are keen on high profitability will carry shorter lines consisting of selected items. Product lines tend to increase over time. Excess manufacturing capacity will put pressure on the product-line managers to develop new items. The sales force and distributors will also pressure for a more complete product line to satisfy their customers.

Line-Stretching Decision

Every company's product-line covers a certain pair of the total range offered by the industry as a whole. Line stretching occurs when a company lengthens its product-line beyond its current range. The company can stretch its line downward, upward or both ways.

Downward Stretch

Many companies initially locate at the high end of the market and subsequently stretch their line downward. For instance, TATA who are the producers of medium and high price/big car segment, stretched downward by entering into small car segment by releasing TATA Indica.

The company is attached at the high end and decides to invade the low end. The company finds that slower growth is taking place at the high end.

The company initially entered the high end to establish a quality image and intended to roll downward.

The company adds a low-end unit to plug a market hole that would otherwise attract a new competitor.

In making a downward stretch the company faces some risks. The new low-end item may cannibalize higher-end items. Or the low-end items might provoke competitors to counteract by moving into the higher end. Or the company's dealers may not be willing or able to handle the lower end products, because they are less profitable or dilute their image.

Upward Stretch

Companies in the lower end of the market might contemplate entering the higher end. They may be attracted by a higher growth rate, higher margins or simply the chance to position themselves as full-line manufacturers. Again, it is Maruti who initially entered in the small car segment entered higher end by production of Maruti 1000 and Maruti Esteem and SUV segment.

An upward decision can be risky. Not only the higher end competitor well entrenched but they may counter attack by entering the lower end of the market. The company's sales representatives and distributors may lack the talent and training to serve the higher end of the market.

Two-way Stretch

Companies in the middle range of the market may decide to stretch their line in both directions.

Line-Filling Decisions

A product line can also be lengthened by adding more items within the present range of the line. There are several motives for line-filling such as reaching for incremental profits; trying to satisfy dealers to complain about lost sales because of missing items in the line; trying to utilize excess capacity; trying to be the leading full-line company and trying to plug holes to keep on competitors. If line-filling is overdone it may result in cannibalization and customer confusion. The company needs to differentiate each item in the consumer's mind. Each item should possess a just noticeable difference. The company should check that the proposed item enjoys more market demand as is not being added simply to satisfy an internal need.

Product Life Cycle

Like human beings, every product has a life span. When a new product is launched in the market, its life starts and the product passes thorough various distinct stages and after the expiration of its life in terms of its capacity to generate sales and profit. This is called Product Life Cycle (PLC).

The Product Life Cycle is an attempt to recognize 'distinct stages' in the 'sales history' of the product. In each stage, there are distinct opportunities and problems with respect to marketing strategy and profit potential. Hence, products require different marketing, financing, manufacturing, purchasing and personnel strategies in the different stages of their life cycle. The PLC concept provides a useful framework for developing effective marketing strategies in different stages of the Product Life Cycle.

There are four stages in the Product Life Cycle – introduction, growth, maturity and decline.

Introduction Stage

The introduction stage starts when the new product is first launched. In this stage only a few consumers will buy the product.

Further, it takes time to fill the dealer pipeline and to make available the product in several markets. Hence, sales will be low a profit will be negative or low. The distribution and promotion expenses will be very high. There are only a few competitors. Regarding pricing, the

Management can pursue either skimming strategy *i.e.* fixing a high price or penetration strategy *i.e.* fixing a low price.

The company might adopt one of several marketing strategies for introducing a new product. It can set a high or low level for each marketing variable, such as price, promotion, distributions and product quality. Considering only price and promotion, for example, management might launch the new product with a high price and lose promotion spending. The high price helps recover as much gross profit per unit as possible which the low promotions spending keeps marketing spending down. Such a strategy makes sense when the market is limited in size, when most consumers in the market know about the product and are willing to pay a high price, and when there is little immediate potential competition.

On the other hand, a company might introduce its new product with a low price and heavy promotion spending. This strategy promises to bring the fastest market penetration and the largest market share. It makes sense when the market is large, potential buyers are price sensitive and unaware of the product, there is strong potential competition and the company's unit manufacturing costs fall with the scale of production and accumulated manufacturing experience.

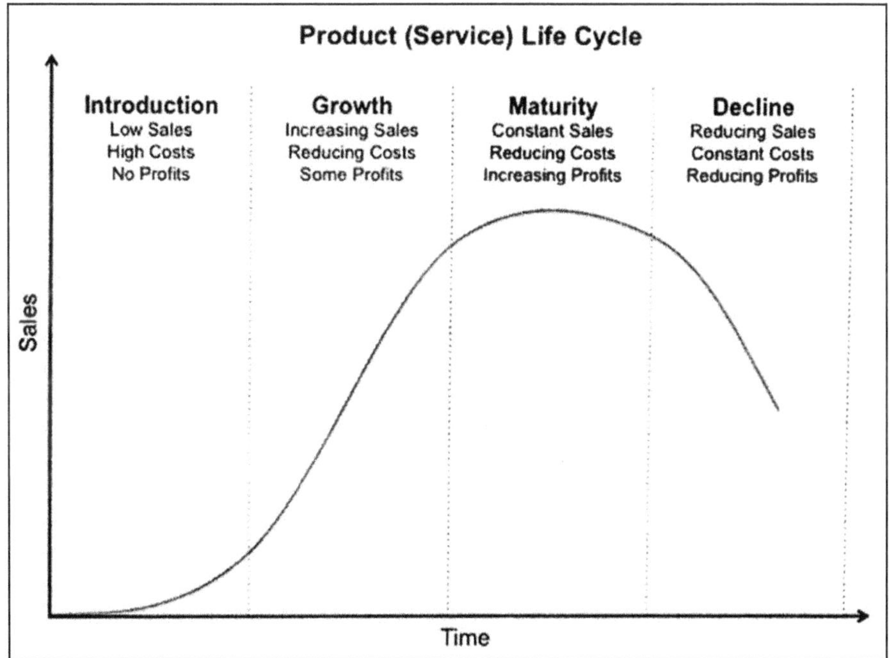

Figure 7.1: Product (Service) Life Cycle.

Growth Stage

If the new product satisfies the market, it will enter a growth stage. This stage is market by quick increase in sales and profits. The early adopters will continue to buy, and later buyers will start following their lead, especially if they hear favourable word of mouth. New competitors enter the market, attracted by the opportunities for high profit. The market will expand. Prices remain the same. Companies maintain their promotional expenditure at the same level or slightly higher level to meet competition and continue educating the market.

During this stage, the company uses the following marketing strategies:

- The company improves product quality and adds new-product features and models.
- It enters new market segments.
- It enters new distribution channel.
- It changes the price at the right time to attract more buyers.

In the growth stage, the firm faces a trade-off between high market share and high current profit. By spending a lot of money on product improvement, promotion and distribution, the company can capture a dominant position. In doing so, it gives up maximum current profit, which it hopes to make up in the next stage.

Maturity Stage

This stage normally lasts longer than the previous stages and it poses strong challenges to marketing management. At this stage, sales will slow down. This stage can be divided into three phases. – growth maturity, stable maturity and decaying maturity.

In the growth maturity phase, the sales start to decline because of distribution saturation. In the stable maturity phase, sales become static because of market saturation. In the decaying maturity phase, the absolute level of sales now starts to decline and customers starts moving toward other products and substitutes. Competitions become acute.

Although many product in the mature stage appear to remain unchanged for long periods, most successful ones are actually evolving to meet changing consumer needs. Product managers should do more than simply ride along with or defend their mature products – a good offense is the best defense. They should consider modifying the market, product and marketing mix.

Marketing Modification: The company should seek to expand the market and enters into new markets. It looks for new users and find ways to increase usage among present customers.

Product Modification: the company should modify the product's characteristics such as quality improvement, features improvement ,style improvement to attract new users and/or usage from current users.

Marketing-mix Modification : The company should also try to stimulate sales through modifying one or more marketing-mix elements such as price cut, step-up sales promotion, change advertisement copy, extending credit *etc.* A major problem with marketing-mix modification is that they are highly imitable by competitors. The firm may not gain as much as expected and in fact all firms may experience profit erosion as they compete each other.

Decline Stage

In this stage, sales decline and eventually dip due to number of reasons including technological advances, consumer changes in tastes and acute competitions. As sales and profit decline some firms withdraw from the market. Those remaining may reduce the number of product offerings.

They may drop smaller market segments and marginal trade channels. They may reduce the promotion budget and prices further. Hence, companies need to pay more attention to their aging products. The firm has to identify those products in the decline stage by regular reviewing sales, market shares, costs and profit trends. Then, management must decide whether to maintain, harvest, or drop each of these declaiming products.

Marketing Strategies during the Decline Stage

Identify the weak products by appointing a product-review committee with representatives from marketing, manufacturing and finance.

The firms may adopt the following strategies.

1. Management may decide to maintain its brand without change in the hope that competitors will leave the industry.
2. Management may harvest by selling whatever is possible in the market.
3. Management may decide to drop the product from the line.

When a company decides to drop a product, the firm can sell or transfer the product to someone else or drop it completely. It must decide to drop the product quickly or slowly. It must decide on how much parts in inventory and service required to maintain service to past consumers.

Uses of PLC Concept

PLC concept's usefulness varies in different decision-making situations. As a planning tool, the PLC concept characteristics the main marketing challenges in each stage and suggests major alternative marketing strategies the firm might pursue. As a control tool, it allows the company to compare product performance against similar products in the past.

Criticism of PLC Concept

1. PLC stages do not have predictable duration. It may vary from product to product.
2. The marketer cannot tell at what stage the product is in as there is no definite line of demarcation between one stage to another stage.

3. Not all products pass through all the stages. It is possible that the product may travel to the first and second stage and die out.
4. A product may not be in an identical stage in all the market segments; it may be in the second stage in one segment, whereas in the third stage in another segment at a particular point of time. Not all products pass through all the stages of its life cycle. Some products are introduced and die quickly; others stay in the mature stage for a long time. Some enter the decline stage and then recycled back into the growth stage through strong promotion or repositioning.

Chapter 8

Pricing Decisions

Among the different components of the marketing-mix, price plays an important role to bring about product-market integration. Price is the only element in the marketing-mix that products revenue.

In the narrowest sense, price is the amount of money charges for a product or service. More broadly, price is the sum of all the values that customer exchange for the benefits of having or using the product or service. Price may be defined as the value of product attributes expressed in monetary terms which a customer pays or is expected to pay in exchange and anticipation of the expected or offered utility.

Pricing helps to establish mutually advantageous economic relationship and facilities the transfer of ownership of goods and services from the company to buyers. The managerial tasks involved in product pricing include establishing the pricing objectives, identifying the price governing factors, ascertaining their relevance and relative importance, determining product value in monetary terms and formulation of price policies and strategies. Thus, pricing play a far greater role in the marketing-mix of a company and significantly contributes to the effectiveness and success of the marketing strategy and success of the firm.

Factors Influencing Pricing

Price is influenced by both internal and external factors. In each of these categories some may be economic factors and some psychological factors; again, some factors may be quantitative and yet others qualitative.

Internal Factors Influencing Pricing

- ☆ Corporate and marketing objectives of the firm. The common objectives are survival, current profit maximization, market-share leadership and product-quality leadership.
- ☆ The image sought by the firm through pricing.

- The desirable market positioning of the firm.
- The characteristics of the product.
- Price elasticity of demand of the product.
- The stage of the product on the product life cycle.
- Turn around rate of the product.
- Costs of manufacturing and marketing.
- Product differentiation practiced by the firm.
- Other elements of marketing mix of the firm and their interaction with pricing.
- Consumption of the product line of the firm.

External Factors Influences Pricing

- Market characteristics
- Buyers behaviors in respect to the given product.
- Bargaining power of the customer.
- Bargaining power of the major suppliers.
- Competitor's pricing policy.
- Government controls/regulation on pricing.
- Other relevant legal aspects
- Social considerations.
- Understanding, if any, reached with price cartels.

Pricing Procedure

The pricing procedure usually involves the following steps:

1. Development of Information Base

The first step in determining the basic price of a company's product(s) is to develop an adequate and up-to-date information base on which price decisions can be based. It is composed of decision-inputs such as cost of production, consumer demand, industry, prices and practices, government regulations.

2. Estimating Sales and Profits

Having developed the information base, management should develop a profile of sales and profit at different price levels in order to ascertain the level assuring maximum sales and profits in a given set of situation. When this information is matched against pricing objectives, management gets the preview of the possible range of the achievement of objectives through price component in the marketing-mix.

3. Anticipation of Competitive Reaction

Pricing in the competitive environment necessitates anticipation of competitive reaction to the price being set. The competition for company's product(s) may arise from similar products, close substitutes. The competitor's reaction may be violent or subdued or even none. Similarly, the reaction may be instant or delay. In order to anticipate such a variety of reactions, it is necessary to collect information about competitors in respect of their production capacity, cost structure, market share and target consumers.

4. Scanning the Internal Environment

Before determining the product price it is also necessary to scan and understand the internal environment of the company. In relation to price the important factors to be considered relate to the production capacity sanctioned, installed and used, the ease of expansion, contracting facilities, input supplies, and the state of labour relations. All these factors influence pricing decisions.

5. Consideration of Marketing-mix Components

Another step in the pricing procedure is to consider the role of other components of the marketing-mix and weigh them in relation to price. In respect of product the degree of perishability and shelf-life, shape the price and its structure; faster the perishability lower is likely to be the price.

6. Selections of Price Policies and Strategies

The next important step in the pricing procedure is the selection of relevant pricing policies and strategies. These policies and strategies provide consistent guidelines and framework for setting as well as varying prices to suit specific market and customer needs.

7. Price Determination

Having taken the above referred steps, management may now be poised for the task of price determination. For determination of price, the management should consider the decisions inputs provided by the information base and develop minimum and maximum price levels. These prices should be matched against the pricing objectives, competitive reactions, government regulations, marketing-mix requirements and the pricing policing and strategies to arrive at a price. However, it is always advisable to test the market validity of its price during test marketing to ascertain its match with consumer expectations.

General Pricing Approaches

Companies set prices by selecting a general pricing approach that includes one or more of the following three approaches:

1. The cost-based approach
   Cost-Plus Pricing
   Break-Even Analysis and Target-Profit Pricing
2. The Buyer-based approach
   Perceive-Value Pricing

3. The Competition based approach
   Going-Rate Pricing
   Sealed-Bid Pricing

1. Cost-Plus Pricing

This is the easiest and the most common method of price setting. In this method, a standard mark up is added to the cost of a product to arrive at its price. For example, the cost of manufacturing a fan is Rs. 1000/- adds 25 per cent mark up and sets the price to the retailer at Rs. 1250/-. The retailer in turn, may mark it up to sell at Rs. 1350/- which is 35 per cent market up on cost. But this method is not logical as it ignores current demand and competition and is not likely to lead to the optimum price. Still mark up price is quite popular for three reasons:

1. Seller have more certainty about costs than about demand and by tying the price to cost, they simplify their pricing task and need not frequently adjust price with change in demand.
2. Where all firms in the industry use this pricing method, their prices will similar and price competition will be minimized to the benefit of all of be them;
3. It is usually felt by many people that cost plus pricing is fairer to buyers as well as to seller.

2. Break-Even Pricing and Target-Profit Pricing

An important cost-oriented pricing method is what is called target-profit pricing under which the company tries to determine the price that would produce the profit it wants to earn. This pricing method uses the popular 'break-even analysis'. According to it, price is determined with the help of a break-even chart. The break-even chart depicts the total cost and total revenue expected at different sales volume. The break-even point on the chart is that when the total revenue equals total cost and the seller neither makes a profit nor incurs any loss. With the help of the break-even chart, a marketer can find out the sales volume that he has to achieve, in order to earn the targeted profit, as also the price that he has to charge for his product.

Buyer-based Approach

Perceive-Value Pricing

Many companies base their price on the products perceived value. They take buyer's perception of value of a product, and not the seller's cost, as the key to pricing. As a result, pricing begins with analyzing consumer needs and value perceptions, and price is set to match consumers' perceived value.

Such companies use the non-price variables in their marketing mix to build up perceived value in the buyer's minds, *e.g.* heavy advertising and promotion to enhance the value of a product in the minds of the buyers. Then they set a high price to capture the perceived value. The success of this pricing method depends on determination of the market's perception of the product's value.

3. Competition-based Approach

Going Rate Pricing

Under this method, the company bases its prices largely on competitor's prices paying less attention to its own costs or demand. The company might charge the same prices as charged by its main competitors, or a slightly higher or lower price than that.

The smaller firms in an industry follow the leading firm in the industry and change their prices when the market leader's prices changes. The marketer thinks that the going price reflects the collective wisdom of the industry.

Sealed-Bid Pricing

This is a competitive oriented pricing, very common in contract businesses where firms bid for jobs. Under it, a contractor bases his price on expectations of how competitors will price rather than on a strict relation to his cost or demand. As the contractor wants to win the contract, he has to price the contract lower than the other contractors. But a bidding firm cannot set its price below costs. If it sets the price much higher than the cost, its chance of getting the contract will be lesser.

Pricing Objectives

A businesses firm will have a number of pricing objectives. Some of them are primary; some of them are secondary; some of them are long-term while others are short-term. However, all pricing objectives emanate from the corporate and marketing objectives of the firm.

Some of the pricing objectives are discussed below:

1. Pricing for a target return.
2. Pricing for market penetration.
3. Pricing for market skimming.
4. Discriminatory pricing
5. Stabilizing pricing.

1. Pricing for a Target Return

This is a common objectives found with most of the established business firms. Here, the objective is to earn a certain rate of Return on Investment (ROI) and the actual price policy is worked out to earn that rate of return. The target is in terms of 'return on investment'. There are companies which set the target at, for example, 20 per cent return on investment after taxes. The target may be for a short-term or a long-term. A firm also may have different targets for its different products but such targets are related to a single overall rate of return target.

2. Pricing for Market Penetration

When companies set a relatively 'low price' on their new product in initial stages hoping to attract a large number of buyers and win a large market-share it is called penetration pricing policy. They are more concerned about growth in sales than in profits. Their main aim is capturing and to gain a strong foothold in the market. This object can work in a highly price sensitive market. It is also done with the presumption that unit cost will decrease when the level of sales reach a certain target. Besides, the lower price may make competitors to stay our. When market share increases considerably, the firm may gradually increase the price.

3. Pricing for Market Skimming

Many companies that launch a new product set 'high prices' initially to skim the market. They set the highest price they can charge given the comparative benefits of their product and the available substitutes. After the initial sales slowdown, they lower the price to attract the next price-sensitive layer of customer.

4. Discriminatory Pricing

Some companies may follow a differential or a discriminatory pricing policy-charging different prices for different customers or allowing different discounts to different buyers.

Discrimination may be practices on the basis or product or place or time. For example, doctors may charge different fees for different patients; railways charge different fares for usual passengers and regular passengers/students. Manufacturers may offer quantity discounts or quote different list prices to bulk-buyers, institutional buyers and small buyers.

5. Stabilizing Pricing

The objective of this pricing policy is to prevent frequent fluctuations in pricing and to fix uniform or stable price for a reasonable period. When price is revised, the new price will be allowed to remain for sufficiently a long period. This pricing policy is adopted, for example, by newspapers and magazines.

New Product Pricing

Pricing a new product is an art. It is one of the most important and dazzling marketing problems faced by a firm. The introduction of a new product may involve some problems in as much as there neither an established market for the product nor a demonstrated demand for it.

The firm may expect a substantial demand for the product though it is yet to be established. Even if there are some near substitutes the actual degree of substitution has to be estimated. Again, there may be no reliable estimate of the direct costs of marketing and manufacturing the product.

Moreover, the cost patterns are likely to change with greater knowledge and increasing volume of production. Yet the basic pricing policy for a new product is the same as for established product – it must cover full in the long run and direct costs in the short run. Of course, there is greater uncertainty about both the demand and costs of the product.

Apart from the problem of estimating the demand for an entirely new product, certain other initial problems likely to be faced are:

1. Discovering a competitive range of price.
2. Investigating probable sales at several possible prices, and
3. Considering the possibility of relation from products substituted by it.

In addition, decisions have to be taken on market targets, design, the promotional strategy and the channels of distribution.

Test marketing can be helpful in deciding the suitable pricing policy. Under test marketing, the product is introduced in selected areas, often at different prices in different areas. These tests will provide the management an idea of the amount and elasticity of the demand for the product, the competition it is likely to face, and the expected sales volume and profits simulation of full-scale production and distribution.

Yet it may provide very useful information for better planning of the full-scale effort. It also permits initial pricing mistakes to be made on small rather than on a large scale.

The next important question is "whether to charge high initial price or a low penetration price".

A High Initial Price (Skimming Price)

A high initial price, together with heavy promotional expenditure,may be used to launch a new product if conditions are appropriate. For example:

(a) Demand is likely to be less price elastic in the early stages than later, since high prices are unlikely to deter pioneering consumers. A new product being a novelty commands a better price.

(b) Is the life of the products promises to be a short one, a high initial price helps in getting as much of it and as fast as possible.

(c) Such a policy can provide the basis for dividing the market into segments to differing elasticities. Bound edition of a book is usually followed by a paper back.

(d) A high initial price may be useful if a high degree of production skill is needed to make the product so that it is difficult and time consuming for competitors to enter on an economical basis.

(e) It is a safe policy where elasticity is not known and the product not yet accepted. High initial price may finance the heavy costs of introducing a new product when uncertainties block the usual sources of capital.

A Low Penetration Price

In certain conditions, it can be successful in expanding market rapidly thereby obtaining larger sales volume and lower unit costs. It is appropriate where:

(a) There is high short-run price elasticity;
(b) There are substantial cost savings from volume production;
(c) The product is acceptable to the mass of consumers;
(d) There is no strong patent protection; and
(e) There is a threat of potential competition so that a big share of the market must be captured quickly.

The objective of low penetration price is to raise barriers against the entry of prospective competitors. Stay-out pricing is appropriate:

i) where total demand is expected to be small. If the most efficient size of the plant is big enough to supply a major portion of the demand, a low-price policy can capture the bulk of the market and successfully holdback low-cost competition.
ii) When potential of sales appears to be great, prices must be set as their long-run level. In such cases, the important potential competitor in a large multi-product firm for whom the product in question is probably marginal. They are normally confident that they can get their costs down to competitor's level if the volume of product is large.

Product-Mix Pricing Strategies

The strategy for setting a project's price often has to be changed when the product is apart of a product mix. In this case, the firm looks for a set of prices that maximizes the profits on the total product mix.

Pricing is difficult because the various products have related demand and costs and face different degrees of completion.

Product-mix Pricing Situations

Product Line Pricing

Companies usually develop product lines rather than single products. In product line pricing, management must decide on the price steps to be set between the various products in a line.

The price steps should take into account cost differences between the products in the line, customer evaluations of their different features, and competitor's prices. if the price difference between two successive products is small, buyers usually will buy the more advanced product.

This will increase company profits if the cost difference is smaller than the price difference. If the price difference is large, however, customers will generally buy the less advanced products.

Optional-Product Pricing

Many companies use optional-product pricing – offering to sell optional or accessory products along with their main product. For example, a car buyer may choose to order power windows, central locking system, and with a CD player. Pricing these options is a sticky problem. Automobile companies have to decide which items to include in the base price and which to offer as options. The economy model was stripped of so many comforts and conveniences that most buyers rejected it. However, General Motors has followed the example of Japanese automakers and included in the sticker price many useful items previously sold only as options. The advertised price now often represents a well-equipped car.

Captive-Product Pricing

Companies that make products that must be used along with a main product are using captive-product pricing. Examples of captive products are razor blades, camera film, and computer software. Producers of the main products (razors, cameras, and computers) often price them low and set high mark ups on the supplies. Thus, Gillete prices its razors slow because it makes its money on the film it sells.

In the case of services, this strategy is called two-part pricing. The price of the service is broken into a fixed fee plus a variable usage rate. Thus, a telephone company charges a monthly rate – the fixed fee – plus charges for calls beyond some minimum number – the variable usage rate.

The service firm must decide how much to charge for the basic service and how much for the variable usage. The fixed amount should be low enough to induce usage of the service, and profit can be made on the variable fees.

By-Product Pricing

In producing agricultural processed products, petroleum products, chemicals and other products, there are often by-products. If the by-products have not value and if getting rid of them is costly, this will affect the pricing of the main product. Using by-product pricing, the manufacturer will seek a market for these by-products and should accept any price that covers more than the cost of storing and delivering them. This practice allows the seller to reduce the main product's price to make it more competitive. By products can even turn out to be profitable for *e.g.* Mollasses in sugarcane processing.

Product-Bundle Pricing

Using product-bundle pricing, sellers often combine several of their products and offer the bundle at a reduced price. Thus computer makers include attractive software packages with their personal computers. Price bundling can promote the sales of products that consumers might not otherwise buy, but the combined price must be low enough to get them to buy the bundle for *e.g.* Insurance bundled with durables.

Price-Adjustment Strategies

Companies usually adjust their basic price for various customer differences and changing situations.

Types of Price-Adjustment Strategies

(1) Discount and Allowance Pricing: Reducing prices to reward customer response such as paying early or promoting the product.

(2) Segment Pricing: Adjustment prices to allow for differences in customers, product, or locating.

(3) Psychological Pricing: Adjusting prices for psychological effort.

(4) Promotional Pricing: Temporarily reducing prices to increase short-run sales.

(5) Value Pricing: Adjusting prices to offer the right combination of quality and service at a fair price.

(6) Geographical Pricing: Adjusting prices to account for the geographic location of customers.

(7) International Pricing: Adjustment prices for international markets.

Discount and Allowance Pricing

Most companies adjust their basic price to reward customers for certain responses, such as early payment of bills, volume purchases and off-season buying. These price adjustments – called discounts andallowances – can take many forms.

A cash discount is a price reduction to buyers who pay their bills promptly. A quantity discount is a price reductions to buyers who buy large volumes. A seasonal discount is a price reduction to buyers who buy merchandise or services out of season.

A trade discount is offered by the seller to trade channel members who perform certain functions, such as selling, storing and record-keeping. Manufacturers may offer different functional discounts to different trade channels because of the varying services they perform.

Allowances are another type of reduction from the list price. Trade allowance is given, for example, or exchange offers. Promotional allowances are payment or price reductions to reward dealers for participating in advertising and sales-support programs.

Segmented Pricing

Companies often adjust their basic prices to allow for differences in customers, products and locations. In segmented pricing, the company sells a product or service at two or more prices, even though the difference in prices is not based on differences in costs. Segmented pricing takes several forms.

Customer-segment pricing : Different customers pay different prices for the same product or service. Railways, for example, charge a concessional fare to children and senior citizens.

Product-form pricing: Different version of the product are priced differently, but not according to differences in their costs.

Location pricing: Different locations are priced differently, even though the cost of offering each location is the same. For instance, theaters vary their seat prices because of audience preferences for certain locations, and state universities charge high tuition fee for foreign students.

Time pricing: Prices vary by the season, the month, the day, and even the hour. Public utilities vary their prices to commercial users by time of day and weekend versus weekday. The telephone company offer slower "off-peak" charges.

Psychological Pricing

Price says something about the product. For example, many consumers use price to judge quality. In using psychological pricing, sellers consider the psychology of prices and not simply the economics.

For example, one study of the relationship between price and quality perceptions of cars found that consumers perceive higher-priced cars having higher quality. By the same token, higher quality cars are perceived to be even higher priced than they actually are.

Another aspect of psychological pricing is reference prices – prices that buyers carry in their minds and refer to when looking at a given product. The reference price might be formed by noting current prices, remembering past prices, or assessing the buying situation. Sellers can influence or use these consumers' reference prices when setting price. For example, a company could display its product next to more expensive ones in order to imply that it belongs in the same class.

Promotional Pricing

With promotional pricing, companies will temporarily price their products below list price and sometimes even below cost. Promotional pricing takes several forms. Supermarkets and departments stores will price a few products as loss leaders to attach customers to the store in the hope that they will buy other items at normal mark-ups. Sellers will also use special event pricing in certain seasons to draw more customers.

Value Pricing

Marketers adopt value pricing strategies – offering just the right combination of quality and good service at a fair price. In many cases, this has involved the introduction of less expensive versions of established, branded products.

Geographical Pricing

A company must also decide how to price its products to customers located in different parts of the country or world. There are five geographical pricing strategies.

1) FOB Pricing: This means the goods are placed free on board of a carrier.
2) Uniform Delivered Pricing: The Company charges the same price plus freight to all customers, regardless of their location.

3) Zone Pricing: All customers within a given zone pay a single total price; the more distant the zone, the higher the price.
4) Basing-point pricing: The seller selects a given city as a "basing point" and charges all customers the freight cost from that city to the customer location, regardless of the city from which the goods actually are shipped.
5) Freight-absorption Pricing: The sellers absorbs all or part of the actual freight charges in order to get the desired business.

International Pricing

Companies that market their products internationally must decide what prices to charge in the different countries in which they operate. In some cases, a company can set a uniform worldwide price.

Administered Price

In real live business situations, product price is not determined as envisaged in the price theory, but is administered by the company's management. An administered or administrative price is set by a company official in contrast to the competitive market prices described in theory. Administered price may, therefore, be defined as the price resulting from managerial decisions of the company. From this, the following characteristics of the administrated price emerge:

(1) Price determination is a conscious and deliberate administrative action rather than a result of the demand and supply interaction.
(2) Administered price is fixed for a period of time or for a series of sale transactions; it does not frequently change.
(3) This price is usually not subject to negotiation; price structure incorporating differentiation; price structure incorporating different variations may, however, be developed to meet specific consumer needs.

The administrative price is set by management after considering all relevant factors impinging on it, *viz.*, cost, demand and competitors' reactions. Since all companies set administrative prices on more or less identical considerations, the prices in respect of similar products available in the market tend to be uniform. The competition, therefore, is based on non-price differentiation through branding, packaging and advertising, *etc.*

Regulated Price

The concept of administrative price may possibly impart a notion that a company is free to fix whatever price if deems fit and buyer have but one choice – either to buy or not to buy. But in real life situation it is not like this. For fear of damages to consumer and national interests, administered prices are subject to state regulation. Therefore, whenever the administered price is set and managed within the state regulation it is termed as regulated price. It may assume two forms. First, the price maybe set by some State agency, say, the Bureau of Industrial Cost a and

Prices or the Tariff Commission and the company just accepts it as given.

Second, the price may be set by a company within the framework or on the basis of the formula given by the State. In India companies, for example, the fertilizer, aluminium and steel industries sell their products at prices fixed by the government, while companies, for example, the cotton textile industry sell products at the price fixed on the basis of a given formula.

In conclusion, it may be said that in the real life Indian business situation it is the 'regulated administrative price' that is relevant for companies and at which products are offered for sale to target consumers.

Pricing over the Product Life Cycle

The price policy can be considered in terms of product life cycle. Anew product, with no competitors has an advantage, and therefore, market skimming policy may be applied. This policy is aimed at getting the 'cream' of the market (the top of the demand curve) at a high before catering to the more price-sensitive segments of the market. In the initial stages the skimming policy can be useful for getting a better understanding of the extent of demand or consumer response as well as to earn adequately to cover the product development costs. The policy may then lead to slow reduction of the price with a view to expand the market. For the new company (as compared with the new product) producing a product in the maturity stage, the preferred object would be penetration pricing – the opposite of skimming. This is unavoidable when the whole demand is elastic and the new entrant company's first aim is to gain entry, a standing or recognition in the market even at a loss for a short period. This policy may also be applied for new product, if the firms expects serious competition very soon after introduction.

Finally, it may be said 'there is not way in which the various factors analyzed earlier can be fed into a computing machine to determine the 'right' price. Individual factors assume varying importance at different times. Basically it is the judgment of the price marker which is the catalytic agent that fuses these various factors into a final decision concerning price. Pricing is an art, not a science. The 'feel' of the market is far more significant than his adeptness with a calculating machine.

Government Control on Pricing

Price controls refer to the Governmental regulations in respect of price fixation.

Usually statutory price control entails imposition of price ceiling so that it does not exceeds consumer capacity to pay. Currently for example, the price of petrol in under statutory price controls. The firms manufacturing these products are assured retention prices which are based on costs, and ensure fair return on investment.

In case of sugar, a dual pricing system has been introduced. Under this system, a manufacturer is required to compulsorily sell a part of its production to the Government at substantially low prices, called levy price.

The rest of production may be sold in the open market as a price the firm deems fit. The statutory price control also envisages the announcement of 'support price' for certain agricultural products like cotton, food grains etc, so as to protect cultivators from price fluctuations.

Voluntary price control envisages formulation of price control measures by the respective industry association under the direction of and according to the guidelines by the Government.

Chapter 9

Distribution Management (Place Decisions)

Channel Management and Physical Distribution

The selection of distribution channels will impinge upon decisions about every other element of the marketing mix. Pricing decisions will be greatly affected by whether the company attempts to mass market through as many wholesale and/or retail outlets as possible, or purposively target a relatively small number of outlets offering its customers high service levels. The amount of promotional effort required of an organisation will be a function of how much, or little, of the selling effort is undertaken by the channels of distribution it uses. The product and/or its packaging may have to be designed to suit the storage and physical handling systems of the distributor.

A distribution channel may be defined as:

"...the set of firms and individuals that take title, or assist in transferring title, to a good or service as it moves from the producer to the final consumer or industrial user."

The importance of channel decisions has not always been recognised. For a long time, marketers only gave thought to appropriate channels of distribution after the product had been developed. However, Bennett claims that:

"... in today's competitive and increasingly global marketplace, managers plan for product distribution as they plan their products."

The same author goes on to state that:

"Modern distribution systems are based on strategic planning, adhere to the marketing concept, focus on target markets, and are consistent and flexible."

Strategic Planning

Distribution channels must be compatible with the strategic marketing plan. If, for instance, a skimming strategy has been adopted or the product requires technical sales support, then mass marketing is probably inappropriate. Alternatively, if large volume sales are required in order to achieve particular profit targets, then selective distribution would be inappropriate.

As new products are introduced, existing channels have to be reassessed since they may not be the right channels for the new product. In some cases, a company will decide not to launch a new product because it does not fit in with existing distribution channels and existing strategy.

Another consideration is the stage of the product's life cycle. It can happen that as the product proceeds through its life cycle the appropriateness of the distribution channel can change. When developing the strategy, thought should be given to how the needs of the product might differ over its life span.

An organisation's distribution strategy is often interconnected with its promotional strategy. As Figure 9.1 illustrates, the distribution system can be depicted as a channel through which products and services move from producer to end user. If the agribusiness concerned believes that its product(s) can be meaningfully differentiated from others on the market, then it may elect to direct the greater part of its promotional effort towards end users. This is termed a *pull strategy*, whereby the objective is to create such a strong preference for the product among end users that the resulting demand pulls the product through the channel of distribution.

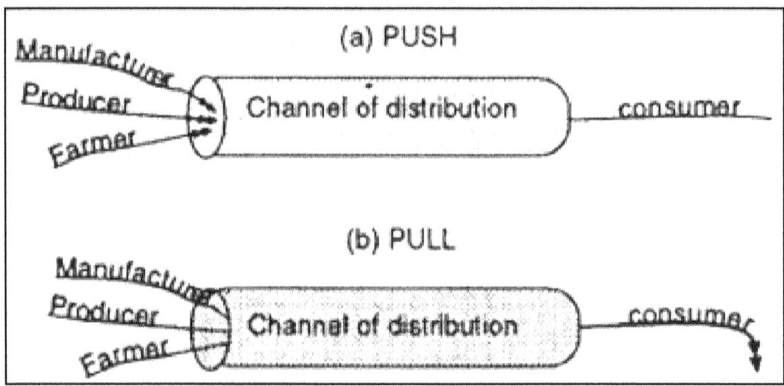

Figure 9.1: Push and Pull Strategies.

Where the product is perceived by end users to be a commodity (or one where there is little difference between brands) then the channel strategy of the agribusiness may be to target much of its promotional effort on intermediaries. If intermediaries can be persuaded to stock the product, in preference to those of competitors, then when customers visit a sales outlet and ask for a product by its generic name it is the product of the company which is supplied. This is termed as *push strategy*. In practice, the promotional strategies of most agribusinesses will be a combination of pulling and pushing the product through the channel of distribution, but there is likely to be more emphasis on one or the other.

Adherence to the Marketing Concept

Agribusinesses which themselves have adopted the marketing concept often experience a problem when their products and services have to be delivered to the end user through intermediaries who are more sales than market-oriented. This should be one of the primary criteria when selecting distributors, *i.e.* the degree of market orientation. In many cases, the producer or supplier will find it difficult to find market oriented intermediaries and in these instances will have to embark on training and education programmes.

Target Marketing

Another important criteria on the selection of distribution channels is the extent to which these focus on the specific market segments that the producer or supplier wishes to penetrate. For example, Agri input dealers distribute a wide range of agricultural inputs to smallholders but do very little business with plantations and estates. Therefore, they would be the wrong type of outlet to handle, say, coconut or tea harvesting equipment since these crops are mainly grown on large estates or plantations.

Thus, it can be seen that channel decisions are central to the organization's overall marketing strategy. Bennett puts it succinctly when he makes the point that:

"Channels are interlocking, highly interdependent, and often complex. Effective distribution is not a patchwork quilt of randomly selected channel members; rather it requires a carefully planned network whose members have clearly assigned functions. The flow of products from manufacturer to wholesaler to retailer to the final buyer depends on systematic, strategic planning and management."

The Value of Middlemen

There are many functions to be carried out in moving the product from producer to customer. Those functions each require funding and, often, specialist knowledge and expertise. Few producers have either the resources or the expertise to carry out all of the necessary functions to get a product/service to the ultimate user. A middleman's remuneration should depend upon the number of marketing functions he/she performs and, more especially, by the efficiency with which they are performed.

The advantages of using middlemen as opposed to marketing direct to end-users can be demonstrated very easily. The efficiency of most marketing systems is improved by the presence of effective intermediaries. This is illustrated in Figures 9.2 a and b. These show that an intermediary between a number of producers and consumers reduces the number of transactions and thereby procurement and selling costs and time are all reduced.

A middleman's existence is justified just so long as he/she performs marketing functions which others cannot or will not, or can perform his/her marketing functions more efficiently than can the producer and/or alternative intermediaries.

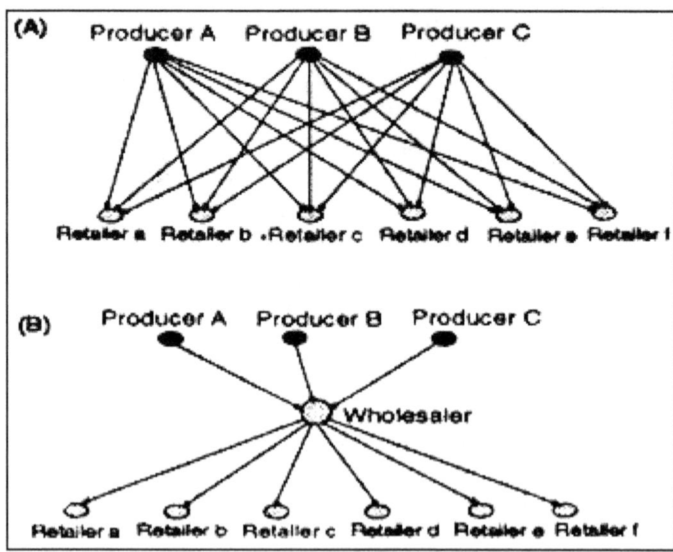

Figure 9.2: A Direct Marketing System and a Marketing System with Intermediaries.

Gaedeke and Tootelian cite three additional reasons why middlemen are commonly employed by producers:

- Intermediaries provide wider market exposure
- Few producers have sufficient capital to market direct and
- Producers can usually earn a higher return on investment by employing available capital in activities other than those of direct marketing.

All too often, in developing countries, middlemen are considered as parasites. The argument made is that it is the producer who, by the sweat of his labour, provides the physical commodity and it is he/she who deserves to gain most from marketing transactions in that product. When it is observed that marketing costs are sometimes four or five times the price paid to the farmer, a sense of injustice can arise. However, the value, if any, that the intermediary adds to the product, by virtue of the functions performed, must be taken into account. McVey states that:

"You can do away with the middleman but you cannot do away with his functions."

In other words, those functions have to be carried out by someone and the expense and risk of doing so has to be met. The real question is not whether middlemen are needed but whether the middleman's remuneration is commensurate with the levels of risk carried and the services provided in the form of marketing functions performed. Furthermore, intermediaries can only be justified if they can perform these functions more efficiently and effectively than the other actual or potential market participants.

Key Decisions in Channel Management

There are a number of key decision areas pertaining to the appointment of intermediaries. These include: price policy, terms and conditions of sale, territorial rights and the definition of responsibilities. In addition, a choice has to be made between extensive and intensive coverage of the market.

Price Policy

List prices, wholesale/retail margins and a schedule of discounts have to be developed. These have to reflect the interests of the intermediary, as well as those of the producer/supplier if lasting alliances are to be formed with channel members.

Terms and Conditions of Sale

In addition to price schedules the producer/supplier must explicitly state payment terms, guarantees and any restrictions on where and how products are to be sold. If the product enjoys a sizeable demand then the producer/supplier may evaluate intermediaries on the basis of performance criteria such as the achievement of sales quota targets, inventory levels, customer delivery times, *etc.* Intermediaries whose performance is below target may have their right to handle the product withdrawn.

Territorial Rights

In the case of certain products, distributors will be given exclusive rights to market a product within a specified territory. This happens, for example, with agricultural equipment. In deciding upon the boundaries of territories the manufacturer or supplier has to strike a balance between defining territories which are sufficiently large to provide good sales potential for distributors but small enough to allow distributors to adequately service the customers within the territory.

Definition of Responsibilities

The respective duties and responsibilities of supplier and distributor have to be clearly defined. For instance, if a customer experiences a problem with a product and requires technical advice or a repair needs to be effected, then it should be immediately clear to both the supplier and the distributor as to which party is responsible for responding to the customer. In the same way, the agreement between the producer/supplier and the distributor should clearly specify which party is responsible for the cost of product training when new employees join the distributor or new products are introduced.

The intensity of distribution *i.e.* the total proportion of the market covered, will depend upon decisions made in the context of the overall marketing strategy. In simple terms there are two alternatives: skimming the market and market penetration. It will be remembered that a skimming strategy involves being highly selective in choosing target customers. Normally, these will be relatively affluent consumers willing and able to pay premium prices for better quality, sometimes highly differentiated, products. It will also be recalled that a penetration strategy is one where the decision has been made to mass market and the object is to make the product available to as many people as possible. The decision as to which of these is adopted as with immediate implications for distribution strategy.

Three principal strategies these being; intensive, selective and exclusive distribution.

Intensive Distribution

Those responsible for the marketing of commodities, and other low unit value products, are, typically, seek distribution, *i.e.* saturation coverage of the market. This is possible where the product is fairly well standardised and requires no particular expertise in its retailing. Mass marketing of this type will almost invariably involve a number of intermediaries because the costs of achieving intensive distribution are enormous. It is simply too expensive in most cases. Where commercial organisations do opt for intensive distribution, channels are usually long and involve several levels of wholesaling as well as other middlemen.

Selective Distribution

Suppliers who appoint a limited number of retailers, or other middlemen, are chosen to handle a product line, have a policy of selective distribution. Limiting the number of intermediaries can help contain the supplier's own marketing costs and at the same time enables the grower/producer to develop closer working relations with intermediaries. The distribution channel is usually relatively short with few or no intermediaries between the producer and the organisation which retails the product to the end user.

Selective distribution is common among new businesses with very limited resources. Their strategy is usually one of concentrating on gaining distribution in the larger cities and towns where the market potential can be exploited at an affordable level of marketing costs. As the company builds up its resource base, it is likely to steadily extend the range of its distribution up to the point where further increases in distribution intensity can no longer be economically justified, it is mostly used with shopping goods.

Exclusive Distribution

Exclusive distribution is an extreme form of selective distribution. That is, the producer grants exclusive right to a wholesaler or retailer to sell in a geographic region. This is not uncommon in the sale of more expensive and complex agricultural equipment like tractors. A Tractor Company, for example, appoints a single dealer to distribute its products within a given geographical area.

Some market coverage may be lost through a policy of exclusive distribution, but this can be offset by the development and maintenance of the image of quality and prestige for the product and by the reduced marketing costs associated with a small number of accounts. In exclusive distribution producers and middlemen work closely in decisions concerning promotion, inventory to be carried by stockists and prices.

Types of Distribution System

Direct Marketing Systems

Where distances between producers and consumers are short, direct transactions between the two groups can take place. Farmers who elect to market their products directly have to trade off the benefits of doing so against the time they are away from farming activities. In the case of industrial markets, direct transactions are common where there are a relatively small number of customers (*e.g.* equipment designed for abattoirs).

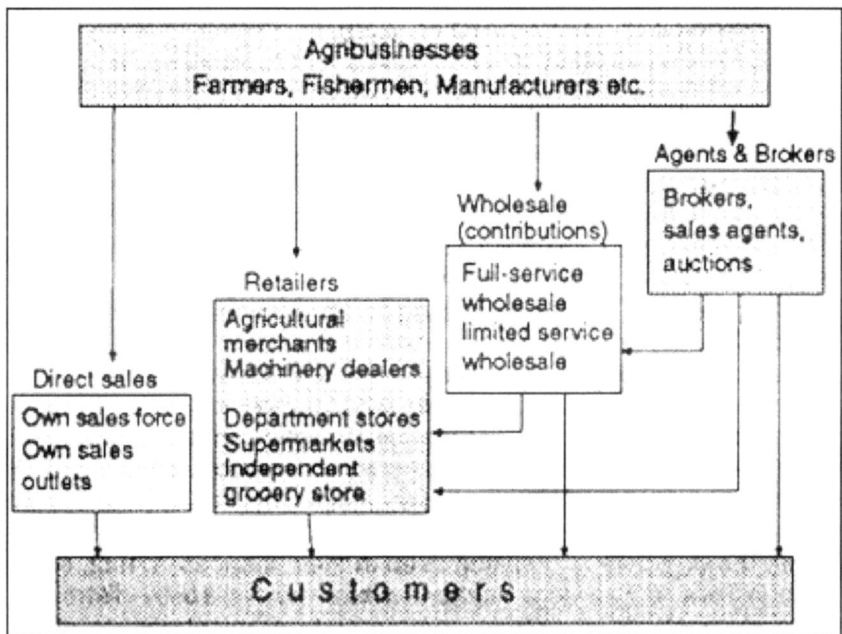

Figure 9.3: The Principal Types of Distribution System.

Retail Institutions

The retail sector includes a wide range of outlets such as merchants, equipment dealers (in the case of farmers), department stores, supermarkets and smaller grocery stores. They are characterised by their dealing with the end user of the product or service.

Wholesalers

Wholesalers make marketing systems more efficient by buying a variety of products, in fairly large quantities, and selling these items on to other businesses who require relatively small quantities of a variety of goods. Wholesalers may service consumer and/or industrial retail outlets. For instance, fruit and vegetable wholesalers often sell to grocery stores (consumer) and hotels, hospitals, schools, prisons, *etc.* (industrial). Some wholesalers offer a full-service *i.e.* they perform all the distribution functions such as selling, pre-delivery inspection (in the case of machinery), technical advice, extension of credit, storage, and delivery. Other wholesalers provide only a limited service.

Sales Agents and Brokers

Sales agents and sales brokers are distinguished from the other types of channel member already described in that they do not take title to the goods. The role of agents and brokers is to facilitate distribution by bringing buyers and sellers together. Sales agents often have close relations with particular growers/processors/manufacturers and contract to sell on their behalf in return for a commission. Some agents negotiate sales for a number of non-competing clients, whilst others handle sales for only one client and usually have the exclusive right to do so, within a specified geographic area. In many respects the sales agent behaves as though he/she were an extension of the client's own sales organisation. Brokers, on the other hand, earn a commission for informing buyers of possible sellers and informing sellers of possible buyers. Clients use the services of a broker intermittently since their supply of the product to the market is intermittent.

Vertical Marketing Systems

This is a system in which the producer(s), wholesaler(s) and retailer(s) act as a unified system. Usually one channel member owns the others, or has contracts with them, or has franchises with others in the channel. The argument for vertically integrated marketing systems is based upon increased efficiency of the system by the removal of duplicated services. They also achieve economies through size, bargaining power and reductions in potential conflicts of interest. In some instances there is a physical coming together of operations and enterprises such as when an abattoir and packhouse physically integrate to provide slaughter, processing, packing and cold store services within a single enterprise. On other occasions the integration has no physical dimension. Examples include:

Franchisers operate by vertically linking several levels of the marketing system. Thus Coca Cola 'wholesales' its syrup concentrate (product) to franchisees who carbonate and bottle and distribute the brand (processing, packaging and physical distribution) to consumers who have been targetted by Coca Cola's heavy advertising (promotion).

In many parts of the world agricultural equipment manufacturers supply machinery (product) to appointed distributors who are given exclusive rights to sell and maintain their whole goods and parts (physical distribution and service) within a specified geographical area. The manufacturer will often provide sales support (promotion) to authorised dealers. In return, these independent distributors agree to abide by the manufacturer's decisions regarding pricing, service levels, stockholding policy and a wide range of other terms and conditions.

Small scale retailers have responded to the competitive advantage of supermarkets by voluntarily integrating their buying and/or wholesaling and marketing operations. This has enabled them to achieve lower operating costs and to offer consumers lower prices than they otherwise could.

Horizontal Marketing Systems

Channels can also develop into horizontal marketing systems in which two or more companies, at the same channel level, cooperate to pursue marketing opportunities. The basis of the marriage is that in combining resources and expertise the partners can achieve some goal that individually they could not. Thus, for example, a seed company and a grain merchant might set up a joint venture to offer farmers a complete package where he buys certified seed from the new enterprise which guarantees to buy his/her grain crop at prevailing market prices.

Strategic alliances of this type are likely to increase in the future. In newly liberalised markets they can be especially useful in protecting local agribusinesses whose low level of capitalisation, outdated technology and inexperience of operating within a competitive marketplace makes them vulnerable when foreign competitors with better resources enter their market.

Marketing to Middlemen

No matter how well a product meets the needs of customers, without effective and efficient distribution it is unlikely to succeed in the marketplace. However, it would be misleading to suggest that producers or suppliers were entirely free to choose which organisations should form the channel for their product (s). In reality, the distribution strategy adopted by most producers more often reflects what is possible rather than what is ideally desired. McVey underlines the point when he states that:

"The middleman is not a hired link in a chain forged by a manufacturer, but rather an independent market, the focus of a large group of customers for whom he buys."

Thus, we are advised by McVey to see all members of the channel we wish to use as customers and we must market *to* them not *through* them.

Power and Conflict in Distribution Channels

Within a distribution channel there is usually a balance of power, and the characteristics of the channel are shaped by the manner in which power is exercised. Sometimes the balance of power in a channel lies with the producer/manufacturer and in other it lies with the intermediary. Moreover, there is always the potential for conflict between channel members.

Conflict between channel members can arise for one or more of the following reasons:

Incompatibility of Goals

Organisations can have conflicting goals. A grower may want to grade the produce in order to achieve a price premium for the top quality produce or to develop a brand image, but the wholesaler may only be interested in selling large volumes of undifferentiated produce.

Confusion over Roles and Rights

For example, a grower may sell part of the produce through local agents and part direct to supermarkets. This may cause conflict because the local agent believes that all sales should go through him/her.

Differences in Perceptions

Among the many potential differences in perceptions, which can result in conflict, are: who the customer is; what the market wants; the objectives of other channel members in participating in the market; and the role which other channel members play in helping the organisation achieve its own objectives.

Whilst manufacturers and producers may think in terms of a distribution system, intermediaries do not necessarily see themselves as part of some other party's 'system', but instead consider themselves as independent operators. If intermediaries do lack a systems orientation, then there are additional prospects of conflict since they will be, naturally, reluctant to compromise their own interests in deference to those of the distribution system as a whole.

Degree of Interdependence

The greater the degree of interdependence between two members of the distribution channel, the greater the potential for conflict. This is because the actions of one directly impinge upon the performance of the other.

Physical Distribution

Gaedeke and Tootelian[10] define physical distribution as:

"...all activities involved in planning, implementing, and controlling the physical flow of raw materials, in-process inventory, and finished goods from point-of-origin to point-of-consumption. The main activities include customer service, inventory control, material handling, transportation, warehousing and storage."

Thus, it is suggested that physical distribution has two components: materials management and marketing logistics. Materials management is concerned with physical supply operations such as procurement and the storage and movement of raw materials to and through processing into a finished product. Marketing logistics deals with the transfer of finished goods to intermediaries, final buyers and end-users.

Physical distribution is often viewed as a necessary support system for the organisation's marketing programme. However, there is an alternative, and more creative perspective which can be taken and that is to see an efficient physical distribution system as a potent marketing tool in its own right, and one which is capable of creating a competitive advantage for the organisation. An organisation which is able, for instance, to supply a wide variety of products speedily at specified times helps reduce the inventory holding costs of the intermediaries being served. Thus the level of interest in the way physical distribution is managed is explained by its potential as a powerful marketing instrument, the opportunity to realise significant savings in marketing costs and by the importance of physical distribution to customer service levels.

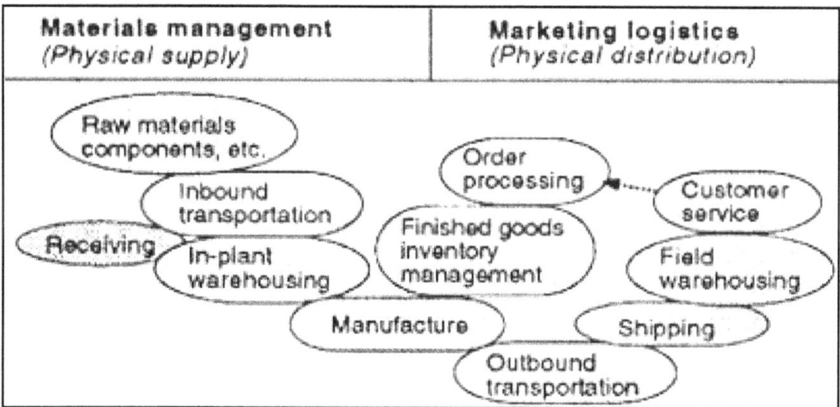

Figure 9.4: The Elements of Business Logistics.

Customer Service Levels

The level of customer service provided by a company is part of the marketing mix. In some instances, a company offers an exceptionally high level of customer service as the principal means of differentiating itself from competitors. Customer service levels are as pertinent to the intermediaries which the agribusiness serves.

For many customers the level of customer service provided by the agribusiness enterprise is as important as any other attribute which it may possess, including the excellence of its products. There are aspects of customer service which have little to do with physical distribution, such as the after-sales service, warranties and the handling of customer complaints, but a large part of customer service is effected through the physical distribution function. A wide range of criteria may be used in evaluating the service level offered by an agribusiness but these are likely to include:

- Timeliness of delivery
- Order size and assortment constraints
- Order cycle time, *i.e.* time interval between order placement and delivery
- Percentage of items out of stock
- Percentage of times an item cannot be supplied from stock (or within a prescribed number of days from order placement)
- Percentage of orders filled accurately
- Percentage of orders arriving in good condition,
- Ease and flexibility of order placement, and
- Competitors' service levels.

Maintaining high levels of customer service carries heavy costs and can only be justified when doing so results in marketing opportunities which otherwise would not be realised. At the same time, the logistics manager must monitor the effects of operating a given level of customer service on profitability. This means considering the trade-offs between the costs involved and the service level offered. It is possible, after all, to provide a level of service above that required or appreciated by the customer.

The Total Distribution Concept

The total distribution concept and the total cost approach are widely applied by managers of physical distribution. They are based on the notion that all elements of physical distribution are so interdependent that a decision made about one element will impact on some or all of the others. Thus, for example, the decision to reduce the number of depots operated by a grain merchant may well reduce costs associated with staffing, wastage, and inventory levels but will also increase transportation costs. The real question is whether the savings in one area match, exceed or fall short of the increased costs in another.

Since, in general, physical distribution managers appreciate that their challenge is to minimise the total costs of the distribution system, rather than the costs of a particular element, they tend to employ the total cost concept. To this end, management must calculate the trade-offs between three categories of cost: transportation costs, order processing costs and stockholding costs. Figure 9.5 shows the general relationship between these different categories of cost.

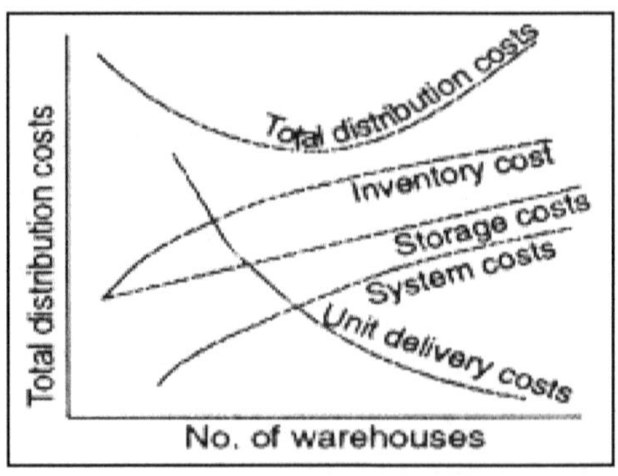

Figure 9.5: The Components of the Total Costs Attached to Physical Distribution.

Storage Costs

Because of economies of scale a large warehouse can be operated at a lower cost than can several smaller warehouses. These economies include the fact that larger warehouses are often better able to achieve better utilisation of space and equipment, overheads incurred in a large warehouse can be spread over a higher throughput of stock items and the amount of money tied up in stock tends to be less for a large depot than for several smaller warehouses. In addition, each separate site will require its own management team and this increases distribution costs further. At some point, however, diseconomies of scale set in and the single central warehouse becomes less attractive in financial terms. This happens, for instance, when depots reach a size where they are difficult to manage and the distances between the warehouse and many of the organisation's customers is so great that transport costs rise to unacceptable levels and the level of service to the customer is adversely affected.

Transportation Costs

The increase in storage costs may be offset, either in whole or in part, by savings made in transportation costs. As the number of warehouses increases, unit transport costs decline due to lower mileages being travelled by delivery vehicles.

For most manufacturers and producers transportation is the major physical distribution cost.

Inventory Carrying Costs

The cost of maintaining sufficient stocks to meet any level of demand is usually prohibitive. Instead, the firm seeks to reach a balance between inventory carrying costs and an acceptable level of customer service.

Among the chief determinants of inventory carrying costs are:

- The greater the number of locations at which stock is held, the greater the level of stocks and carrying costs.
- Longer order cycles result in higher stocks, and vice versa and
- As the product portfolio increases so does the amount invested in stocks

With respect to the effect of increasing the number of warehouses located in various areas, this would be, as was said earlier, to increase stock holding costs.

System Costs

The last category of the costs are those termed system costs. These include costs attached to order processing activities, the maintenance of information systems and communications between sites.

Total Distribution Cost

Changes in one element of the distribution system can have a dramatic, and often unexpected effect on other elements of the system and upon the system as a whole. Hence the need to view the physical distribution system as a whole. Total distribution costs analysis can be used to this end. By bringing together the various types of distribution cost, the effects of proposed changes in one area of distribution can be assessed in terms of their impact on other individual elements and upon the system as a whole.

Warehouse Management

The functions of warehouses are to provide cost-effective storage, in suitable conditions, for the organisation's products and materials. The existence of a warehouse is justified by the extent to which it contributes to the efficiency and effectiveness of physical distribution functions. The geographical location of a warehouse should be determined by production sites and the physical position of target markets.

Warehouse managers have a number of important challenges including:

- determining the most appropriate unit load(s)
- optimising space utilisation
- reducing the movement of labour, equipment and products/materials to a minimum
- establish a safe, secure warehouse environment and
- keeping costs to a minimum.

Inventory Management

The management of inventory can have a major impact upon the profitability of an organisation. If inventory levels are too low then there is the risk of stockouts, *i.e.* the inability to meet an order. This can result in a loss of revenues, profits and customer goodwill. On the other hand, if the inventory levels are too high then the organisation can experience cash flow problems since so much of its capital is tied up in stocks. When inventory levels are high then there is also an increased risk of spoilage, pilferage and obsolescence.

Summary

Channel decisions are integral to the strategic marketing plan. Distribution systems should adhere to the marketing concept, focus on target markets and have sufficient flexibility to enable an organisation to respond to market changes and new market opportunities.

A middleman's remuneration should depend upon the number of marketing functions he/she performs and, more especially, by the efficiency with which they are performed. The efficiency of most marketing systems is improved by the presence of effective intermediaries. Whilst it always possible to by-pass or remove intermediaries from a marketing system, the functions which they performed in the past will remain to be performed, as will their costs.

Chapter 10

Promotion Decisions

Marketing Communications

Marketing communications are intended to both inform and persuade a target audience, with a view to influencing the behaviour of that group. The behaviour of interest to agribusinesses can range from encouraging farmers to adopt improved husbandry practices or to grow a particular crop (or variety of crop), to encouraging industrial or consumer buyers to try a product or service. Each element of the marketing mix must be designed so as to further the overall marketing strategy, and this includes marketing communications.

The Nature of Marketing Communications

In fact, without effective marketing communications the consumer remain unaware of products and services they need, who might supply them and the benefits which both product and suppliers can offer. Moreover, it is impossible to develop effective and efficient marketing systems without first establishing channels of communication. Even the best products do not sell themselves. Marketing communications serve five key objectives:

- ☆ the provision of information
- ☆ the stimulation of demand
- ☆ differentiating the product or service
- ☆ underlining the product's value,
- ☆ regulating sales.

Marketing communications takes four forms - advertising, sales promotion, personal selling and publicity. These must be formulated within a co-ordinated marketing communications plan. If there is more than one target market then there will need to be more than one communications programme. Like all other elements of the marketing mix, it must be tuned to the characteristics and needs of the target market.

Advertising

Advertising is the most visible element of the communications mix because it makes use of the mass media, *i.e.* newspapers, television, radio, magazines, bus hoardings and billboards. Mass consumption and geographically dispersed markets make advertising particularly appropriate for products that rely on sending the same promotional message to large audiences. Many of the objectives of advertising are only realised in the longer term and therefore it is largely a strategic marketing tool.

Sales Promotion

Sales promotion employs short-term incentives, such as free gifts, money-off coupons, product samples *etc.*, and its effects also tend to be short-term. Therefore, sales promotion is a tactical marketing instrument. Sales promotions may be targetted either at consumers or members of the channel of distribution, or both.

Public Relations

Public relations is an organisation's communications with its various publics. These publics include customers, suppliers, stockholders (shareholders, financial institutions and others with money invested in the business), employees, the government and the general public. In the past, organisations thought in terms of publicity rather than public relations. The distinction between advertising and publicity was based on whether or not payment was made to convey information via the mass media. Advertising requires payment by the sponsor of the message or information whilst publicity is information which the media decides to broadcast because it is considered newsworthy and therefore no payment is received by the media from a sponsor. It is more common these days to speak of public relations than of publicity. Public relations is much more focused in its purposes.

The objectives of public relations tend to be broader than those of other components of promotional strategy. It is concerned with the prestige and image of the organisation as a whole among groups whose attitudes and behaviour can impact upon the performance and aims of the organisation. To the extent that public relations is ever used in product promotion, it constitutes an indirect approach to promoting an organizations products and/or services.

Personal Selling

This can be described as an interpersonal influence process involving an agribusiness' promotional presentation conducted on a person-to-person basis with the prospective buyer. It is used in both consumer and industrial marketing and is the dominant form of marketing communication in the case of the latter.

Developing an Appropriate Communications Programme

Marketing strategy is derived from an organisation's corporate strategy. The marketing strategy then has to be translated into a strategic plan, or set of strategic plans if the organisation intends to exploit opportunities in more than one target market. Strategic plans are very broad statements of principles which the organisation believes will lead it to achieve its marketing objectives within a chosen target market. These principles become operational when they are expressed in the form of a marketing plan consisting of a detailed blueprint for each element of the marketing mix product, distribution, pricing and marketing communications.

The framework in Figure 10.1 shows the connection between marketing communications and the marketing strategy. It also highlights the main stages involved in developing a marketing communications programme..

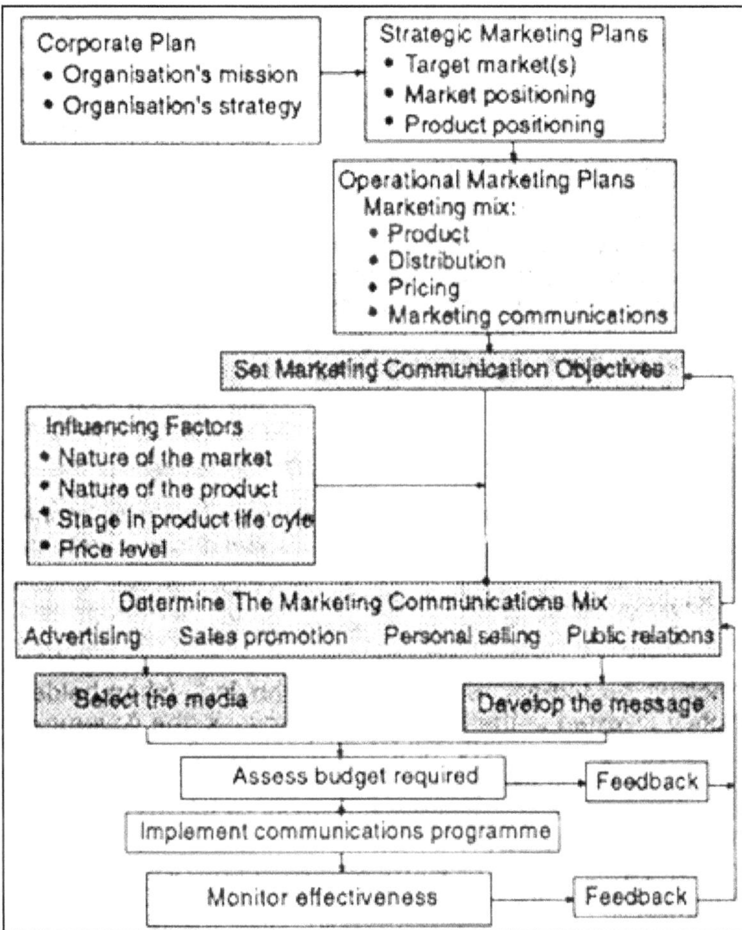

Figure 10.1: Developing a Marketing Communications Programme.

Once the overall marketing strategy has been determined and the marketing plan has been outlined, it is necessary to develop a set of operational communication objectives. It is only when this is done that an appropriate marketing communications mix can be designed. There are, however, a number of intervening factors to be considered before the communications mix is finalised. These include the nature of both the product and the market, the stage at which the product lies in its life cycle and the relative value of the product in terms of its price to potential purchasers. Having decided upon the communications mix, the promotional message can be determined and the medium or media best suited to delivering this message can be chosen. At this point, the budgetary implications of the decisions made so far have to be considered. If the cost of the communications programme exceeds the resources available to the organisation, then there may have to be an adjustment in the communications mix. In some instances, the organisation may conclude that it can adjust the communications mix to reduce the cost to an affordable level but that the revised communications package is unlikely to achieve the original objectives. Faced with this situation, the organisation may resort to revising its marketing communications objectives. Once the budget has been set the programme can be implemented. The effectiveness of the programme has to be measured against its objectives and, if necessary, adjustments or wholesale revisions of the programme will be made.

Setting Marketing Communication Objectives

A three step approach is proposed and this takes into account the longer term outcomes of marketing communications. The three steps are:

1. Identify the target segment
2. Determine the behavioural change to be brought about
3. Decide what needs to be done to bring about the change in behaviour.

1. Identifying the Target Segment

The identification of the target audience is obtained from the marketing strategy and marketing plan. There may, however, need to be a refinement of the target group for a particular promotional campaign. There is a direct relationship between the degree of precision with which the target group is defined and the clarity with which communication objectives can be stated. Moreover, if the target group is defined with precision this greatly assists in deciding upon both the content of the promotional message and the medium chosen to carry it.

2. Intended Behavioural Changes

There should be a clear understanding of what behavioural changes the communications programme is intended to bring about. Is it: To increase usage among existing customers? To convert non-users to users?To establish new uses for an existing products? To reduce the amount of brand switching and encourage more users to be brand loyal? To enable customers to make better, more effective, more efficient or less wasteful use of the product and thereby increase its value to them? It is possible to measure the extent to which changes in behaviour have occurred, but marketing communication objectives can only become operational when the intended behavioural changes are stated with precision and without ambiguity.

3. Deciding what Needs to be Done

The third step in developing operation objectives for marketing communications, is to specify the required course of action. To increase the number of uses of a product might only require an awareness campaign, to improve the way in which a product is used (*e.g.* farmer's application of plant growth chemicals) would probably involve an educational campaign, to create a liking of the product a programme aimed at attitudinal change would be necessary, and the conversion of non-users of the product to users is likely to focus upon creating a conviction about its benefits and attributes.

Factors Influencing the Communications Mix

There are at least 5 major influences on what makes a given mix of promotional techniques appropriate. These are: the nature of the market, the nature of the product, the stage in the product life cycle, price and the funds available for promotional activities.

a) Nature of the Market

An organisation's target audience greatly influences the form of communication to be used. Where a market is comprised of relatively few buyers, in reasonable proximity to one another, then personal selling may prove efficient as well as effective. Conversely, large and dispersed markets are perhaps unsuitable for personal selling because the costs per contact will be high. The customer type also has an impact. A target market made up of industrial purchasers, wholesalers or retailers is more likely to be served by organisations which employ personal selling than is a market of consumers.

Another important consideration is the state of the prospective customer's knowledge and preferences with respect to the product or service. In some cases, the task will be to make potential customers aware of a product which is entirely new to them, whilst in others, the aim will be to attract them away from a competing product. The two tasks are quite different in nature and may require the use of differing forms of communication.

b) Nature of the Product

Highly standardised products, with minimal servicing requirements, are less likely to depend upon personal selling than are custom designed products that are technically complex and/or require frequent servicing. Standardised, high sales volume products, especially consumer products, will probably rely more on advertising through the mass media. Where the product is targetted at a narrow market segment or where those who can use the product effectively are few in number then personal selling will prove the more cost effective method of communications.

c) Stage in the Product Life Cycle

The promotional mix must be matched to a product's stage in the product life cycle. During the introductory stage, heavy emphasis is placed upon personal selling to convey the attributes and benefits of the product. Intermediaries are personally contacted to increase awareness, interest and, if possible, commitment to the product. Trade shows and demonstrations are also frequently used to inform and educate prospective dealers/retailers and, sometimes, consumers. Advertising at this stage is chiefly informative, and sales promotion techniques, such as product samples and money-off coupons, are designed to achieve the goals of getting potential customers to try the product.

As a product graduates into the growth and maturity stages, advertising places greater emphasis upon persuasion, with the ultimate objective of encouraging the target market to become purchasers of the product. Personal selling efforts continue to be directed at marketing intermediaries in an attempt to expand distribution. As more competitors enter the market, advertising begins to stress product differences to establish brand loyalty. Reminder advertisements begin to appear in the maturity and early decline stages.

Thus, we see that as a product progresses through the product life cycle, both the marketing objectives, and the promotional mix used to achieve them, may well change.

d) Price

The fourth factor impinging upon the promotional mix is that of price. Advertising and/or sales promotion are the dominant promotional tools for low unit value products due to the high per contact costs in personal selling. Higher value products can justify, and usually require, personal selling.

e) Promotional Budget

A real barrier to implementing any promotional strategy is the size of the promotional budget. Mass media advertising tends to be expensive although the message can reach large numbers of people and hence the cost per contact is relatively low. For many new or smaller firms the costs are prohibitive and they are forced to seek less efficient but cheaper methods. Ideally, a promotional strategy should first be developed and then costed rather than designing a promotional strategy around a preset budget.

The Table 10.1 summarises the main influences upon the selection of the elements of the communications mix.

Table 10.1: Choosing between Personal Selling and Mass Media

	Personal Selling	*Mass Media*
Market		
Number of buyers	Few	Many
Geographic	Concentrated	Dispersed
Type of market	Industrial	Consumer
Product		
Product complexity	Custom	Standardised
Service level required	High	Low
Life cycle stage	Introductory to early growth	Maturity to early stage of decline
Pricing	High unit value	Low value

The Marketing Communications Mix

The next set of decisions is to determine the role of each element of the promotional mix. Depending upon the situation, it is likely that more emphasis will be given to certain forms of promotion than to others. Table 10.2 provides a brief overview of the main advantages and disadvantages of each element of the promotional mix.

Table 10.2: The Main Promotional Methods

Form of Promotion	*Advantages*	*Disadvantages*
Personal selling	Permits flexible presentation and gains immediate response.	Costs more than all other forms per contact. Difficult to attract good sales personnel.
Sales promotion	Gains attention and has instant effect.	Easy for others to imitate
Advertising	Appropriate for reaching mass audiences. Allows direct appeal and control over the message.	Considerable waste. Hard to demonstrate product. Hard to close sale. Difficult to measure results.
Public relations	Has a high degree of believability when done well.	Not as easily controlled as other forms. Difficult to demonstrate or measure results.

Advertising

Advertising is characterised as a form of communication which its sponsor pays to have transmitted via mass media such as television, radio, cinema screens, newspapers, magazines and direct mail. It is intended to both inform and persuade. Lancaster and Massingham describe advertising as being:

"...concerned with the identification and presentation of desirable and believable benefits to the target audience in the most cost effective way."

In some respects the aims of corporate communications would seem more a responsibility of public relations rather than advertising. However, reference here is made to the use of medium for which the organisation has paid. Public relations does not pay to make use of the mass media.

Since the effects of advertising are only evident in the longer term it should be treated as a strategic rather than tactical tool of the marketing communications mix. Advertising does not have the immediate impact of creating a customer. Instead, it has a hierarchy of effects as depicted in Figure 10.2. Lavidge and Steiner's hierarchy of effects model describes communication as a process rather than a simple outcome in the form of a sale.

Awareness

Consider the task facing a government which is attempting to persuade farmers in a frequently drought-stricken area to switch some of their production from maize to more drought-resistant sorghum. The initial step is for advertising to create an awareness of both the economic and technical benefits of sorghum which would accrue to farmers within drought-stricken areas. There may also be an awareness

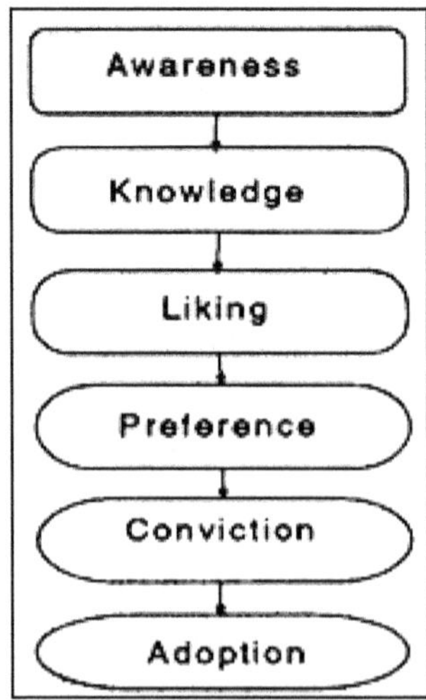

Figure 10.2: The Hierarchy of Effects Model.

task to be accomplished with respect to new sorghum varieties whose higher yields help compensate for the superior economic rewards of growing maize in a good season. Levels of awareness can be measured and thereby used as a measure of the effectiveness of advertising. For example, prior to beginning a planned advertising campaign a target such as the following might be set:

'Within 3 months of the campaign running, we expect at least 30 percent of farmers in region X to be aware of the new sorghum variety and to be able to recall the 3 main technical benefits that are claimed for the variety in the campaign.'

Subsequent research among the region's farmers would permit management to determine whether the advertising had accomplished this target or not.

Knowledge

The next step is to instill, in farmers, a given level of knowledge about, for instance, how to choose economically viable sorghum varieties and the best husbandry practices to maximise yields; and economic results, the technical and commercial benefits of the new variety and how these are achieved. It is unlikely that advertising alone can communicate this type of information. The technical nature of the information would suggest that farmers would wish to put questions to sales personnel and/or extension agents in order to obtain further explanation.

Whatever combination of marketing communications is used, quantitative targets can again be set and the performance of the programme can be evaluated against them. It is particularly important that post-campaign research establishes the level of understanding among the target group. It should never be assumed that just because a message is received it is also understood.

Once an awareness and understanding has been built up among the target audience the marketer can then focus on establishing a liking or positive attitude towards the crop. This might be done, for instance, by promoting the virtues of the new variety, *e.g.* drought-resistant, high-yielding and palatable.

Preference

Even though the campaign may create a positive predisposition towards the product or service, the product may not be preferred to the alternatives. In the case of the hypothetical new sorghum variety, the target audience may like what it hears about the variety but this may not yet be preferred to existing varieties or to planting maize. Preference can be created by promoting the *comparative* advantages of the new product or service over its alternatives. To create preference the promotional message must convey benefits which alternatives do not possess.

Conviction

It is possible that whilst the target audience has developed a preference for a product or service their conviction about that product or service is not yet strong enough to actually cause them to adopt it. Here, the role of communication is to convince the target audience that the claimed benefits of the product or service are both real and sufficiently great to warrant a change in their behaviour. For example, prospective growers of the new sorghum variety will want to see the benefits for themselves through field trails and demonstration plots, and will perhaps want to converse with farmers who have already grown the new variety. This hypothetical example indicates that the medium of communication (*e.g.* printed media or demonstrations) and sources of information (*e.g.* extension personnel or other farmers) may change from stage to stage.

Adoption

The final step is for the target audience to adopt the crop, husbandry practice, product or service. The original hierarchy of effects model had purchase as its final step but here the term adoption is preferred because it emphasises that the ultimate objective of promotion is to encourage a long term change in behaviour and not a one-off trial or purchase. To facilitate the initial purchase or trial of the product or service the promotional campaign might centre around a low introductory price or enable potential customers to try it on a limited basis. Prospective growers of the new sorghum variety could be offered seed at a discounted price or the seed might be specially packed in small sample sachets so that it could be sown on a trial plot of land.

In summary, what needs to be recognised is that it is unlikely that all of the steps in the communication process can be accomplished by a single advertisement or advertising campaign.

Sales Promotion

In contrast to advertising, sales promotion is more tactical than strategic. It is usually applied to create an immediate impact, but one which is unlikely to be sustained in the longer term. Thus, marketers tend to use promotion to address short term problems such as reducing the cash burden of overstocked products, stimulating demand during what is traditionally the low season, selling off stocks which are becoming obsolete or are likely to spoil if they remain in storage. Sales promotions may be targetted at consumers, industrial buyers (*e.g.* crop processors or food manufacturers), channel intermediaries (*e.g.* traders, wholesalers or retailers) or the organisation's own sales force.

Table 10.4: Types of Consumer Sales Promotion

Sales Promotions Targetted on Customers	
Type of Promotion	*Examples*
Discount coupons or money-off packs	Discounts on the full price encourage product trial, *e.g.* Rs.5 off the regular price that will apply to a new pesticide.
Premiums	Products offered free or at a discount act as an incentive to buy a related product, *e.g.* farmers buying 25 litres or more of a new pesticide get a 5 litre pack of herbicide free.
Lotteries, games or competitions	Intended to create interest and excitement among customers, *e.g.* farmers may be offered the opportunity to win a knapsack or tractor mounted agrochemical sprayer.
Samples	Free samples encourage product trial, *e.g.* farmers could be given a small pack of pesticide and invited to apply it to a small plot and compare results either with a plot to which no pesticide has been applied or against a competing brand.
Point-of-sale merchandising	These specially designed display units and literature are intended to create impulse (*i.e.* unplanned) purchases. They are located close to the place where the customer pays for the goods or service, *e.g.* the packs of pesticide could be arranged on an attractive rack displaying the manufacturer's name and situated, close to the checkout in a farmers service centre.

Table 10.4 gives examples of typical forms which sales promotion takes. Many of these forms are equally applicable in consumer and industrial markets.

Sales promotions may be targetted on intermediaries as well, or instead of, consumers. Many types of promotion are used in both sectors. Sometimes, however, their objectives are slightly different. Table 10.5 describes the main forms of trade promotions and their various purposes.

Public Relations

Publicity and public relations are not one and the same thing. Organisations often seek publicity, *i.e.* to disseminate newsworthy items of information about itself, its products/services or about its personnel through the media but does not pay to do so as in the case of advertising. Instead, the organisation hopes that the item is sufficiently newsworthy to appear in an editorial feature, in a newspaper or magazine, or that a radio and/or television station will want to interview an official of the organisation about the item.

Table 10.5: Types of Trade Sales Promotions

Sales Promotions Targetted on Trade Channels	
Type of Promotion	*Examples*
Trade allowances	These temporary price reductions are intended to be passed on, in whole or in part, to the end customer. Thus, intermediaries can elect to have a higher margin per unit or higher volume sales.
Bonus purchases	An agricultural merchant may be offered 24 packs of pesticide for the price of 20. Such bonuses are not intended to be passed on to customers but are an incentive for the merchant to increase the order size.
Competitions	These are directed at the sales and/or service personnel of intermediaries and if sponsored by a manufacturer/grower are intended as an incentive to place particular emphasis on selling that supplier's products or services., *e.g.* a salesman achieving total orders in excess of 1,000 litres of pesticide might win a cash prize.
Cash incentives	Cash bonuses paid to a middleman's sales personnel can help push the product through the channel of distribution.
Cooperative advertising/ promotions	Suppliers and middlemen sometimes share the cost of an advertising campaign or promotion, *e.g.* An agricultural merchant wishing to run a local campaign may obtain assistance from one or other of his/ her main suppliers.
Trade shows and exhibitions	An industry's trade association may organise fairs and exhibitions which offer its members the opportunity to communicate with a well defined target audience. Both manufacturers and intermediaries may participate in these events.

Publicity can be a highly effective communication tool, since 'news' is often perceived by the target group to have greater authenticity and credibility than 'advertising'. Moreover, it can penetrate the defences of individuals who intentionally ignore advertising and sales personnel. The main disadvantage of publicity is that the organisation has relatively little control over it.

Public relations may be defined as:

"...the deliberate, planned and sustained effort to establish and maintain mutual understanding between an organisation and its public."

The 'public' referred to in this definition is any group having an actual or potential interest in, or impact upon, an organisation's prospects of achieving its goals. Such publics would be:

The Community

An organisation needs to be accepted by the local community. To this end, a community relations programme should be established. Such a programme should devise ways for the organisation to become involved in community activities. A public relations programme can give an organisation a 'personality' and, hopefully, one which the local community likes.

Consumers

Public relations should be used to nurture a positive image of the organisation and its products and services, a belief in its intrinsic fairness in dealings with customers and the perception that the organisation values loyal customers.

Other Channel Members

Wherever the organisation is placed within the marketing channel (as a grower, processor, wholesaler, retailer *etc.*) it should take cognisance of the need to develop and maintain positive relations with its partners within the marketing system. The public relations programme should make them feel like partners, *e.g.* by making them privy to privileged information about the organisation's products, promotional programmes, marketing plans, future developments and/or policies.

Opinion Leaders

Pressure groups and trade associations are examples of groups which can influence both public and government opinion and therefore should be a target for the organisation's public relations activities.

Government

The lobbying of politicians is a sensitive issue but in most countries around the world it is accepted as a reality. Public relations programmes should be designed to create a two-way flow of communications between industry and government (or between a trade association, such as a farmers' union, and government). That is, the organisation should be creating a positive predisposition towards it whilst it should be receiving advance information, from government, on matters such as proposed legislative changes that could impact upon its activities.

Financial Institutions

Bankers, finance brokers, investment analysts and other lending institutions are an important public for all commercial organisations. They need to have confidence in the financial stability and prospects for growth since directly or indirectly they will affect the organisation's access to capital. Public relations programmes targetted at this group are therefore very important.

Media

Sound press relations can give an organisation access to the 'news' channel of communication through which it can disseminate positive information to all of its publics. Through its public relations programme, the press should be given a ready response to all reasonable requests for information within the limits of commercial confidentiality, that the organisation is candid about its intentions and actions.

Employees

Organisations must recognise the need to 'market' themselves to their own employees as much as to other publics. Internal public relations often suffers from neglect. The loyalty and commitment of employees to the organisation and its goals cannot be taken for granted. An internal public relations programme can also help build an understanding between the organisation and its personnel as well as helping develop an enduring trust between them.

The methods employed by public relations professionals include:

- ✰ Open days
- ✰ Sponsorship
- ✰ In-house publications
- ✰ Community projects
- ✰ Press releases
- ✰ Video films
- ✰ Training courses,
- ✰ Annual reports.

Public relations has perhaps a different but complementary role to that of other forms of communication. It will be most effective, and controllable, when it has specific objectives, with respect to specific publics, and when it is coordinated with the forms of marketing communication.

Personal Selling

Personal selling complements both advertising and sales promotion. Many organisations have a sales force comprised of a number of representatives who have face-to-face contact with the customer. The division of responsibilities between sales representatives may be based on geographical areas or on product groups. For instance, an agrochemical company could divide the market into geographic regions and assign a representative to each district. He or she would have responsibility for selling all of the company's products to the assigned area. Alternatively, the same agrochemical company could organise its sales force so that representatives handle either animal health products or crop protection products. This would make sense if, within a country, farming tended to be specialised into arable and livestock, whereas it would perhaps be less appropriate if mixed farming is the norm and two representatives, from the same firm, were calling on the same farmer.

Reid defines personal selling as:

"...the process of analysing potential customers' needs and wants and assisting them in discovering how such needs and wants can best be satisfied by the purchase of a specific product, service or idea."

Sales representatives have at least 7 key tasks:

1. **Prospecting**: Sales representatives find and develop business with new customers.
2. **Communicating**: Sales representatives communicate information about the organisation's philosophy, produce/products and/or services and communicate needs. preferences and problems which customers have and the organisation can meet or resolve.

3. **Selling**: Sales representatives should be trained in the art of selling approaching, presenting, countering objections, closing sales and nurturing a long-term customer relationship.
4. **Servicing**: Sales representatives provide various services to customers, such as helping resolve their problems with his/her own organisation, rendering technical assistance, arranging financing and expediting delivery.
5. **Information gathering**: Sales representatives carry out market research and intelligence work and complete visit reports. Representatives are able to collect information on competitor activity as well as the future needs of customers.
6. **Complementing advertising**: The activities of sales representatives should complement other elements of the promotional mix. The sales approach has to be consistent with the selling propositions conveyed through advertising and sales promotion. Where possible, customer visits should be timed to coordinate with the other promotional mix components.
7. **Allocating**: Sales representatives are able to evaluate the value of various customers to the organisation and advise on the allocation of scarce produce/products at times of shortage.

Thus, we observe that whilst selling is of fundamental importance, the sales representative has a number of other vital objectives, but at core he/she is part the organisation's promotional effort and is an important contributor to marketing communications.

Training the Sales Force

Organisations which send their newly hired sales representatives immediately into the field are almost invariably disappointed by the results. It is true that training programmes can be expensive. Trainers have to be hired, materials purchased and, perhaps, facilities have to be rented. Moreover, the organisations are paying people who are not yet selling. There are also opportunity costs. Experienced sales representatives have to be withdrawn from the field for on-going training, and sales opportunities may be foregone. However, training, and re-training, is necessary if the sales force is to be effective. The sales manager's task is to ensure that training costs are outweighed by the value added to the company's business by having a better equipped sales force.

The great majority of sales personnel need to be trained to become active sellers as opposed to being passive order takers.

Sales force training programmes have several goals, including:

1. **Sales representatives need to understand and identify with the company:** The first part of most sales training programmes is devoted to describing the company's history and objectives, the organisation and lines of authority, the chief officers, the company's financial structure and

facilities, and the chief products and sales volume, as well as the current and prospective customer base.

2. **Sales representatives need product knowledge**: Sales trainees are shown how the products are produced and how they function in various uses and in different environments.
3. **Sales representatives need to understand customers' needs, buying motives, and buying habits**: They need to learn about the organisation's and competitors' strategies and policies.
4. **Sales representatives need to know how to make effective sales presentations**: Sales representatives require training in the principles of salesmanship. In addition, the company outlines the major sales arguments for each product, and some provide a sales script.
5. **Sales representatives need to understand field procedures and responsibilities**: Sales representatives learn how to divide their time between existing and prospective customers; how to prepare reports, organise their schedules and select efficient routes.
6. **Sales representatives need to understand their role in marketing intelligence gathering**: Some individuals are quicker than others to realise that they have a role in market intelligence gathering. Indeed, representatives repeatedly fail to report information collected in the course of their work because they do not appreciate that it constitutes marketing intelligence. Hence, the need for training in this area.

Training programmes have to be evaluated against the performance (and/or improved performance) of the sales force. Quantitative evidence might include increased sales turnover and sales volumes, larger average order sizes, increases in new accounts, a decline in customer complaints and returns, lower levels of absenteeism, *etc.*

Developing the Message

In most instances, the attention of the target audience can only be held for a relatively short time. That is, the potential customer will spend only a matter of seconds, or at most minutes, reading an advertisement in the printed media, will spare a limited time conversing with sales personnel or extension agents and will quickly lose interest in broadcast messages when these are perceived to be too long. Thus marketers must be selective in the points of information they seek to communicate. Whilst a product or service could have a large set of selling points, these will have to be narrowed down to a select few. Moreover, the single most important selling point will be the one to be included in the principal slogan or headline. This is sometimes termed the unique selling proposition (USP). A USP should only be decided upon after customer research has determined meaningful and important messages (*e.g.* there is little to be gained from promoting the nutritional superiority of hammer milled whole grain over roller milled refined grain when consumers believe the latter is superior in taste, colour and texture).

Selecting the Media

The media plan has to be developed in concert with the overall marketing communications strategy. The hierarchy of effects model, described earlier in this chapter, stressed the multiple stages through which the target customer must be brought and that different media might be more successful at some stages than others. Therefore, it is likely that a mix of media will have to be used within a single marketing communications programme.

Criteria for selecting communications media include:

- ☆ level of exposure
- ☆ level of impact
- ☆ nature of the target audience
- ☆ cost and cost effectiveness.

Message Exposure

Marketers are interested in the potential number of message exposures that a given medium offers. The total level of exposure is a function of **reach** and **frequency**. Reach is the number of people exposed to the message. For example, to the extent that a higher percentage of rural populations, in developing countries, have access to radio as opposed to television, radio will have the greater reach for this target audience. Frequency is the average number of times an individual is exposed to the message. If the target audience were say farmers, who tend to read a newspaper 2 – 3 times per week but listen to the farming news, on radio, 7 days per week, then radio is likely to achieve the higher frequency rating.

Invariably, there is a trade-off between reach and frequency. Communications budgets will stretch only so far and so more spent on one will reduce the amount that can be spent on the other.

Impact of the Promotional Message

It can be argued that the impact of a promotion has more to do with the message than the medium. Nonetheless, the medium is an influencing factor on the levels of awareness, comprehension, believability and retention. Radio, being a purely audio medium, will be limited in its impact on farmers' levels of understanding of the operation of a piece of agricultural equipment that is new to them. Visual communication would be important in this case. In the same way, the retention of information is generally higher for audio-visual communications than it is when the information is presented only in audio form.

The Target Audience

Media have to be selected according to their ability to reach the target audience. This involves analysing the demographic structure of the market socio-economic groups, age groups, language, ethnic groups, *etc.* Thereafter, marketers can seek to identify media that reach the target group(s) for *e.g.*, regional channels to reach rural farmers.

Cost and Cost Effectiveness

Some forms of media may prove too expensive for a particular communications budget and although these may have great potential in reaching the target audience, they will be unavailable. Even when this is not the case, it is incumbent upon the marketer to identify the most cost effective media.

The cost-per-thousand method (CPM) is one of the most commonly used in measuring the cost effectiveness of promotional media.

Establishing the Promotional Budget

Deciding upon the amount to be spent on promotion is one of the most challenging tasks marketing managers have to face. There are simply no scientific solutions to the problem. Since no one has ever established a mathematical relationship between promotional expenditures and their effects, either in terms of sales volumes or revenues, there is no universally accepted formula for setting the promotional budget. Instead, a number of pragmatic approaches have been established over the years. The main budget setting methods are percentage-of-sales, fixed-sum-per-unit, competitive parity, residual-sum and objective-and-task.

Percentage-of-Sales

The method involves setting the budget as a percentage of either last year's sales or forecasted sales for next year. Thus, brands or products which are performing well get additional promotional support. The popularity of this approach is probably due to its simplicity. However, it suffers from several weaknesses, for example high sales volumes do not necessarily reflect high profitability, there is little support for marketing managers wanting to turn 'problem' products into 'rising stars' and when budgets are set according to forecasted sales there is motivation to inflate sales estimates. Another problem with this approach is that using percentages of sales leads to sales determining the promotional mix.

Fixed-Sum-per-Unit

Some organisations elect to set a specified amount for each unit sold or produced.

Competitive Parity

This approach is simply one of keeping pace with immediate competitors. The organisation will try to work out approximate expenditure levels by two or three close competitors and will then seek to match those expenditures. It represents a reactive or defensive approach to promotional budget setting. The method also discourages organisations from taking a more aggressive marketing stance by seeking to gain a competitive advantage.

Residual-sum

This is a term for allotting what the organisation perceives itself to be able to afford after all other budgets have been set. The danger is that in years of good business there may be over-budgeting whilst in times of low sales, when demand most needs to be stimulated, the amount available for promotion falls.

Objective-and-Task

An organisation employing the objective-task approach will first specify its communication objectives and will then estimate how much it will cost to achieve those objectives. This is the approach to promotional budget setting recommended in this text. It has the benefit of encouraging marketing managers to set specific communication goals. When these are not attained the communications mix can be reevaluated and modified.

Perhaps the fundamental weakness of this budgeting method is the implied assumption of cause-and-effect. That is, there is an assumption of a direct relationship between promotion and marketing performance but as has already been said, other elements of the marketing mix will impact upon sales, as will many uncontrollable exogenous factors.

Whichever approach to setting the promotional budget is chosen it should be recognised that it has been established on a less than optimal basis.

Monitoring the Effectiveness of Marketing Communications

The last step in the development of a marketing communications programme is to evaluate the effectiveness of the programme. The evaluation has two components: communications effects and market performance.

Communications Effects

Research into communications effects involves the evaluation of a single advertisement. Research in this area focuses upon measuring variables such as attention levels, message comprehension, message retention and intention to purchase. This type of research is often termed copy research. Both broadcast and printed promotional material can be evaluated. Whilst the techniques employed differ in their detail they essential involve exposing a sample of people drawn from the intended target group and exposing them to the proposed advertisements or promotions. For example, a printed advertisement can be inserted in a dummy magazine and given to the sample. After a suitable period of time these people are asked to recall the advertisements seen and to report as much of the detail of the content of the ads selling propositions, images, applications, *etc.* In the same way, an audience can be recruited to watch television programmes with trial advertisements inserted at the beginning of the programme(s) and/or in the commercial intervals and/or at the end of the programmes. They too can be questioned about the content of the advertisements and the impressions that they made upon the audience.

Market Performance

To a limited extent and in certain situations, the effects of promotion on sales can be measured. The effects of special offers and coupons can be measured by redemption rates. Two approaches which are widely pursued in industrialised countries are as follows:

1. **Field experiments**: The organisation selects two geographical areas which are similar in terms of socio-economic groups, levels of disposable income *etc.* and launches a promotional campaign in one area but not the other. After a period of time, sales in the two areas are compared. The assumption is that the only difference between the two areas is the absence or presence of the promotional campaign and so differences in sales are explained by the promotional campaign.
2. **Analysis of historical market data**: Promotional expenditures and sales data can be compared using mathematical or econometric models to first describe the relationship between sales and promotion. Where these can be successfully described there is the prospect of developing other models capable of predicting sales, given a certain level of promotional effort.

The first of these approaches requires the application of very strict controls and careful matching of the areas or markets to be compared. The second approach requires good quality data. That is, the data must be detailed, precise, free from error and must extend over a considerable period of time.

Summary

The establishment of effective communication channels between sellers and buyers is a prerequisite of success in agricultural marketing. Marketing communications serve to both inform and persuade. More specifically, through the promotional mix, advertising, sales promotion, personal selling and public relations organisations can provide information to other market participants, stimulate demand, differentiate products and services, underline a product's value and regulate sales.

Marketing communication objectives are derived from the marketing plan and must be consistent with the other elements of the marketing mix. These objectives must be operational and require identification of a target market, a specification of any desired changes in that target group's behaviour and a set of performance targets.

Index

T

U

V

W

www.ingramcontent.com/pod-product-compliance
Ingram Content Group UK Ltd.
Pitfield, Milton Keynes, MK11 3LW, UK
UKHW021953270726
14060UKWH00002B/501